Low Frequency Radio Astronomy and the LOFAR Observatory

Astrophysics and Space Science Library

More information about this series at http://www.springer.com/series/5664

George Heald • John McKean • Roberto Pizzo
Editors

Low Frequency Radio Astronomy and the LOFAR Observatory

Lectures from the Third LOFAR Data Processing School

Springer

Editors
George Heald
CSIRO Astronomy and Space Science
Kensington
WA, Australia

John McKean
Kapteyn Astronomical Institute
University of Groningen
Groningen, The Netherlands

ASTRON
Dwingeloo, The Netherlands

Roberto Pizzo
ASTRON
Dwingeloo, The Netherlands

ISSN 0067-0057 ISSN 2214-7985 (electronic)
Astrophysics and Space Science Library
ISBN 978-3-030-13241-5 ISBN 978-3-319-23434-2 (eBook)
https://doi.org/10.1007/978-3-319-23434-2

Cover illustration: Background: LOFAR superterp photograph (c) Top-Foto, Assen (The Netherlands). Inset images were created during the Third LOFAR Data Processing School by the participants. Left inset: MSSS image courtesy of Hoang Ngoc Duy & Edwin Retana Montenegro. Right inset: Long-baseline image courtesy of Colm Coughlan & Rachel Ainsworth

This Springer imprint is published by the registered company Springer Nature Switzerland AG
The registered company address is: Gewerbestrasse 11, 6330 Cham, Switzerland

This book is dedicated to the memory of Professor A.G. (Ger) de Bruyn (1948–2017). Ger played an essential role in the development of LOFAR, and he was central to enhancing the scientific impact of the telescope. Ger also had a gift for educating and inspiring young radio astronomers, and so it is particularly fitting that this volume should be dedicated to his memory. Indeed several of the lecturers and editors of this book were taught and inspired by Ger. We hope that some portion of his knowledge and spirit can be found in these pages.

Preface

The LOw Frequency ARray (LOFAR) is a low-frequency radio interferometer centred in the northern part of the Netherlands. LOFAR's aperture array design, huge physical extent and demanding computational requirements all contribute to a challenging (but rewarding!) experience for research astronomers. To educate the LOFAR user community, and in particular the students and postdocs that personally experience most of the challenges and rewards, a series of LOFAR Data Processing Schools was initiated. These Schools are organised and hosted by ASTRON, the Netherlands Institute for Radio Astronomy. ASTRON built and operates the LOFAR telescope.

The Third LOFAR Data Processing School (of a continuing series) took place on 17–21 November 2014. It featured nineteen lectures by experts in the field and with LOFAR in particular; five tutorial sessions; two evening lectures; a tour of the LOFAR core area near Exloo; and several tasty group dinners. Fifty participants took part in the School. The School website can be found at http://www.astron.nl/lofarschool2014/ and includes photos from during the School, lecture slides and a full list of the participants.

This book, based on material from the Third LOFAR Data Processing School, was inspired by the excellent and ubiquitous standard reference *Synthesis Imaging in Radio Astronomy II* (1999), Taylor, Carilli and Perley eds., or *SIRA-II* for short. That book comprises a collection of lectures from the Sixth NRAO/NMIMT Synthesis Imaging Summer School in Socorro, New Mexico (17–23 June 1998). *SIRA-II* is one of a small number of essential volumes that should always be at hand during the course of a radio astronomer's education. Our aim with this book was not to replace *SIRA-II*, but to provide additional information that is specifically needed to supplement the education of young radio astronomers working at low frequencies, and especially with LOFAR.

This book was a long time in the making, and many aspects of the LOFAR system and data processing have changed in the meantime. Indeed many features of the LOFAR telescope and software continue to change at a healthy pace. Although we have done our best to include up-to-date information at the time of publication, with that continual spirit of improvement in mind we encourage the reader to refer to the LOFAR webpages to stay informed about the latest system and software status.

Kensington, WA, Australia George Heald
Dwingeloo, The Netherlands Roberto Pizzo
Groningen, The Netherlands John McKean
March 2018

Data Processing School participants enjoying a tour of the LOFAR core near Exloo, The Netherlands.

Acknowledgements

The editors would like to acknowledge the many people without whom you would not be reading this book.

First, many thanks to the local organisers of the Third LOFAR Data Processing School for their tireless efforts to keep the School running smoothly, for organising the dinners, and generally fostering a fun and comfortable environment for the students. The LOC included Roberto Pizzo, Liesbet Elpenhof, Wilfred Frieswijk, Marco Iacobelli, Fritz Möller and Marjan Tibbe.

Second, we are grateful to the lecturers and tutors for contributing their time and wisdom to the preparation of the School, and for inspiring the participants with their enthusiasm and love of low-frequency radio astronomy and LOFAR. The lecturers (and evening speakers) were George Heald, Michiel Brentjens, Menne J. Norden, Roberto Pizzo, Tammo Jan Dijkema, Jan David Mol, John McKean, Ger de Bruyn, Elizabeth Mahony, Maaijke Mevius, Michael Wise, Bas van der Tol, Javier Moldón, Raymond Oonk, Jason Hessels, Stijn Buitink, Richard Fallows and Wilfred Frieswijk. The tutorial session leaders were Tammo Jan Dijkema, Hanno Holties, Elizabeth Mahony, Emanuela Orrú and Javier Moldón.

Special thanks go to the authors of the chapters included in this book, not only for committing to contributing their material but also for their patience as the entirety of the book came together.

Last, but certainly not least, our thanks go out to the participants of the Third LOFAR Data Processing School! In addition to working very hard to generate the image on the cover of this book, your questions and thirst for knowledge inspired the lecturers and tutors to do their best in helping you navigate the tricky world of LOFAR. We hope that you find this book useful.

Contents

Part I Introduction to Low-Frequency Radio Astronomy and the LOFAR Telescope

1 Introduction to Low Frequency Radio Astronomy 3
George Heald
1.1 Historical Context 3
1.1.1 Scientific and Technical Background 5
1.1.2 A History of Radio Astronomy 6
1.1.3 Review of Radio Interferometry 9
1.2 Low Frequency Science 11
1.2.1 Cosmic Magnetism 12
1.2.2 Epoch of Reionization 12
1.2.3 High Energy Cosmic Rays 13
1.2.4 Solar Physics and Space Weather 13
1.2.5 Surveys 13
1.2.6 Transients 14
1.2.7 Other Topics 14
1.3 Low Frequency Radio Facilities 14
References 16

2 LOFAR Overview 19
Michiel A. Brentjens and Jan David Mol
2.1 Introduction 19
2.2 Antennas 20
2.3 Observing Bands 21
2.4 Configuration 22
2.5 Signal Path and Data Flow 25
2.6 COBALT: Central Signal Processing 26
2.6.1 Correlator Processing Pipeline 27
2.6.2 Tied Array Beam Former Processing Pipeline 28

2.7 Phased Array Beam Forming 29
2.7.1 Beam Patterns 30
2.7.2 Grating Responses 31
2.7.3 A Hierarchy of Beams 33
2.8 Summary 35
References 36

3 LOFAR Station Processing 37
Menne J. Norden
3.1 Antennas 37
3.1.1 LBA 38
3.1.2 HBA 39
3.2 Stations 41
3.3 Receiver Unit 43
3.3.1 HBA Control Modem 44
3.3.2 RCU Modes 45
3.4 Digital Processing 46
3.4.1 Beamlet Processor (BLP) 47
3.4.2 Subband Statistics (SST) 48
3.4.3 Beamlet Statistics (BST) 48
3.4.4 Crosslet Statistics (XST) 49
3.5 Station Clock 49
References 51

Part II Fundamentals of LOFAR Data Processing

4 RFI Flagging, Demixing and Visibilities Compression 55
Tammo Jan Dijkema
4.1 Best Discretization in Time and Frequency 55
4.2 Default Preprocessing Pipeline: DPPP 56
4.3 Flagging RFI 57
4.3.1 Inspecting Data to Spot RFI 58
4.3.2 Flagging Ranges of Data (preflagger) 59
4.3.3 Automatic Flagging Using AOFlagger 60
4.4 Removal of Bright Sources 60
4.4.1 Smart Demix 61
4.4.2 Averaging Data with DPPP 62
References 63

5 Calibration of LOFAR 65
John McKean and Ger de Bruyn
5.1 Introduction 65
5.2 The Evolving Playing Field for Imaging and Calibration 70
5.3 From Calibration to Self-Calibration 71
5.3.1 The Self-Calibration Philosophy 71
5.3.2 Calibration Regimes and Unknowns 74

5.4 Flux-Density and Spectral Bandpass Calibration 79
5.5 Calibration Procedure 80
5.5.1 Black Board Self-Calibration: An Introduction 81
5.5.2 Specifying the Sky Model 83
5.5.3 Calibration Strategies 84
5.5.4 Inspecting and Flagging Solutions 85
5.5.5 Direction Dependent Calibration 86
5.6 Advanced Calibration Techniques 87
5.6.1 A Facet Calibration Approach 87
5.6.2 A Global Solve Using SAGECal 89
5.7 Summary 90
References 90

6 Error Recognition in LOFAR Data 93
Elizabeth Mahony
6.1 Diagnosing Errors in the u,v Plane Compared to the Image Plane 93
6.2 Recognising Calibration Errors in the Image Plane 94
6.3 Recognising Imaging Errors 94
6.4 Wide-Field Imaging Effects 96
6.4.1 *w*-Projection and *A*-Projection Effects 96
6.4.2 Bandwidth-Smearing and Time-Average Smearing 97
6.4.3 Sidelobes from Bright Sources in the Field 97
6.4.4 Direction Dependent Effects 99
6.5 Can You Do Science with It? 99
References 101

7 Ionospheric Effects 103
Maaijke Mevius
7.1 Introduction 103
7.1.1 Electromagnetic Propagation 105
7.2 Dispersive Delay 106
7.2.1 Separation of Phase Effects 107
7.2.2 Direction Dependent Correction 110
7.3 Faraday Rotation 111
7.3.1 Differential Faraday Rotation 112
7.3.2 RM Correction 113
References 115

8 Wide Field Imaging 117
Sebastiaan van der Tol
8.1 Introduction 117
8.2 Data Model 118
8.3 Imaging as a Data Fitting Problem 120

8.4 Gridding and Degridding 122
8.4.1 Sampling of the Convolution Function 128
8.5 W-Projection 130
8.6 W-Stack Algorithm 131
8.7 A-Projection 132
8.8 Estimating the Effect of the Gridding Parameters on the Final Image 133
References 137

9 The LOFAR Standard Imaging Pipeline 139
George Heald
9.1 Context 139
9.2 Goals and Assumptions 140
9.3 SIP Strategy, Framework, and Components 141
9.3.1 Preprocessing Pipeline 141
9.3.2 Calibration and Imaging Pipelines 144
9.3.3 Self-Calibration Cycle 150
9.4 Using the SIP 151
9.5 Limitations and Future Work 152
References 154

Part III Advanced Topics in LOFAR Data Processing

10 Polarization Imaging with LOFAR 159
Michiel A. Brentjens
10.1 Introduction 159
10.2 Polarized Electromagnetic Waves 160
10.3 Stokes Parameters 161
10.4 Stokes Visibilities 163
10.5 Jones Matrices 165
10.6 Faraday Rotation 167
10.7 Ionosphere 171
10.8 Antenna Beam Polarization 174
10.9 Summary 177
References 178

11 Long Baseline Imaging with LOFAR 179
Javier Moldón and Eskil Varenius
11.1 Introduction 179
11.2 The International LOFAR Stations 180
11.2.1 Sampling of Fourier Space 181
11.2.2 Resolution and Sensitivity 181
11.2.3 Field of View 183
11.3 Calibration of International LOFAR Stations 185
11.3.1 Phase Calibration Using International Stations 186
11.3.2 Amplitude Calibration of International Stations 192

11.4 Observing Strategy 193
11.5 Practical Considerations 195
11.5.1 Optimising Use of Available Bandwidth 196
11.5.2 Finding Calibrators 197
11.5.3 Form a Combined Station 198
11.5.4 Getting LOFAR Data into AIPS 198
References 199

12 Spectral Line Data Analysis with LOFAR 201
J. B. Raymond Oonk
12.1 Introduction 201
12.1.1 Radio Recombination Lines 202
12.2 Spectral Capabilities of LOFAR 202
12.3 Spectral Line Calibration 203
12.3.1 Correcting the Bandpass 206
12.3.2 Continuum Subtraction 207
12.3.3 Doppler Correction 208
12.3.4 Flagging Statistics 208
12.3.5 Processing and Storage Requirements 209
12.4 Concluding Remarks 209
References 210

13 Particle Physics with LOFAR 211
Stijn Buitink
13.1 The Transient Buffer Boards 211
13.1.1 Description of Buffer Boards 211
13.1.2 Reading Out the TBBs 212
13.2 Calibration 213
13.2.1 Relative Gain Calibration 213
13.2.2 Absolute Gain Calibration 214
13.2.3 The Antenna Model 215
13.3 Working with Time Series Data 215
13.3.1 RFI Cleaning 216
13.3.2 Pulse Finding 217
13.3.3 Fitting Pulse Arrival Directions 218
13.3.4 Pulse Polarization 218
13.4 Introduction to Cosmic Ray Analysis 219
13.4.1 Radio Emission from Air Showers 219
13.4.2 Air Shower Reconstruction with LOFAR 220
References 222

14 High Time Resolution with LOFAR 225
Jason Hessels and Richard Fallows
14.1 Scope of This Chapter 225
14.2 Motivation for High Time Resolution with LOFAR 226

14.3 Observing at High Time Resolution with LOFAR 228
14.3.1 Brief Summary of Beam Definitions 228
14.3.2 Standard Beam-Formed Modes 229
14.3.3 Customizing the Beam-Formed Modes 232
14.3.4 Other Observing Modes 233
14.3.5 Observing Challenges 233
14.3.6 Planning an Observation 236
14.4 Analyzing LOFAR Beam-Formed Data 237
14.5 The LOFAR Standard Pulsar Pipeline (PulP) 241
14.6 The Dynamic Spectrum Toolkit 241
14.7 Future Prospects 242
References 243

Index 245

Astronomical Objects and Fields Index 251

Contributors

Michiel A. Brentjens ASTRON, Dwingeloo, The Netherlands

Ger de Bruyn ASTRON, Dwingeloo, The Netherlands

Stijn Buitink Vrije Universiteit Brussel, Ixelles, Belgium

Tammo Jan Dijkema ASTRON, Dwingeloo, The Netherlands

Richard Fallows ASTRON, The Netherlands Institute for Radio Astronomy, Dwingeloo, The Netherlands

George Heald ASTRON, Dwingeloo, The Netherlands

CSIRO Astronomy and Space Science, Kensington, WA, Australia

Jason Hessels ASTRON, The Netherlands Institute for Radio Astronomy, Dwingeloo, The Netherlands

Anton Pannekoek Institute for Astronomy, University of Amsterdam, Amsterdam, The Netherlands

Elizabeth Mahony ASTRON, Dwingeloo, The Netherlands

Sydney Institute for Astronomy, School of Physics A2, The University of Sydney, Camperdown, NSW, Australia

John McKean ASTRON, Dwingeloo, The Netherlands

Kapteyn Astronomical Institute, University of Groningen, Groningen, The Netherlands

Maaijke Mevius ASTRON, Dwingeloo, The Netherlands

Jan David Mol ASTRON, Dwingeloo, The Netherlands

Javier Moldón Jodrell Bank Centre for Astrophysics, School of Physics and Astronomy, The University of Manchester, Manchester, UK

ASTRON, Dwingeloo, The Netherlands

Menne J. Norden ASTRON, Dwingeloo, The Netherlands

J. B. Raymond Oonk ASTRON, Dwingeloo, The Netherlands

Leiden Observatory, Leiden University, Leiden, The Netherlands

Sebastiaan van der Tol ASTRON, Dwingeloo, The Netherlands

Eskil Varenius Onsala Space Observatory, Department of Earth and Space Sciences, Chalmers University of Technology, Onsala, Sweden

Acronyms

AARTFAAC	Amsterdam-ASTRON Radio Transients Facility and Analysis Center
ADC	Analog to Digital Converter
ADI	Alternating Direction Implicit
AGN	Active Galactic Nucleus
AIPS	Astronomical Image Processing System
AP	Antenna Processor
APERTIF	Aperture Tile In Focus
ATCA	Australia Telescope Compact Array
BAO	Baryon Acoustic Oscillation
BBS	BlackBoard Self-calibration
BF	Beamformer
BLP	Beamlet Processor
BP	Board Processor
BS	Block Synchronisation
BST	Beamlet Statistics
CASA	Common Astronomy Software Applications
CCU	Central Control Unit
CEP	Central Processing
CIT	Centre for Information Technology
CITT	Calibration and Imaging Tiger Team
COBALT	Correlator and Beam former Application for the LOFAR Telescope
CODE	Centre for Orbit Determination in Europe
CR	Cosmic Ray
CS	Core Station
CS	Coherent Stokes Beam
CV	Complex Voltage
DAL	Data Access Library
DFT	Direct Fourier Transform
DPPP	Default Pre-Processing Pipeline
DTC	Dynamic spectrum Toolkit Container

EMU	Evolutionary Map of the Universe
EoR	Epoch of Reionization
EPS	Expanded Polystyrene
FE	Fly's Eye (beam)
FFT	Fast Fourier Transform
FIFO	First In First Out
FIR	Finite Impulse Response
FPGA	Field Programmable Gate Array
FRB	Fast Radio Burst
FWHM	Full Width at Half Maximum
GBT	Green Bank Telescope
GLEAM	Galactic and Extragalactic All-sky Murchison Widefield Array (survey)
GMRT	Giant Metrewave Radio Telescope
GPS	Global Positioning System
GPU	Graphical Processing Unit
GSM	Global Sky Model
HBA	High Band Antenna
HBA-FE	High Band Antenna Front End
HI	Neutral Hydrogen
I^2C	Inter-Integrated Circuit
IAU	International Astronomical Union
ICM	Intracluster Medium
IDG	Image Domain Gridding
IGM	Intergalactic Medium
ILT	International LOFAR Telescope
IS	Incoherent Stokes Beam
ISM	Interstellar Medium
KSP	Key Science Project
LBA	Low Band Antenna
LBCS	LOFAR Long-Baseline Calibrator Survey
LCASS	LOFAR Cassiopeia A Spectral Survey
LCP	Left Circular Polarization
LCU	Local Control Unit
LEP	Left Elliptical Polarization
LM	Levenberg-Marquardt
LNA	Low Noise Amplifier
LOFAR	Low Frequency Array
LORA	LOFAR Radboud air shower Array
LoTSS	LOFAR Two-metre Sky Survey
LSM	Local Sky Model
LST	Local Sidereal Time
LTA	Long Term Archive
LTE	Local Thermodynamic Equilibrium
LVDS	Low Voltage Differential Signalling

LWA	Long Wavelength Array
MAC	Monitoring and Control
MFS	Multifrequency Synthesis
MKSP	Magnetism Key Science Project
MoM	Management of Measurements
MRO	Murchison Radio-astronomy Observatory
MS	Measurement Set
MSSS	Multifrequency Snapshot Sky Survey
MWA	Murchison Widefield Array
NDPPP	New Default Preprocessing Pipeline
NVSS	NRAO VLA Sky Survey
OLAP	Online Application Processing
PAPER	Precision Array for Probing the Epoch of Reionization
PCB	Printed Circuit Board
PFS	Pre Filter Structure
PFT	Pipelined FFT
P/O	Processing/Observing
PPF	Poly Phase Filter
PPS	Pulse Per Second
PSF	Point Spread Function
PSWF	Prolate Spheroidal Wave Function
PulP	Pulsar Pipeline
PWM	Pulse Width Modulator
PyBDSF	Python Blob Detector and Source Finder
RAD	Ring Adder
RCP	Right Circular Polarization
RCU	Receiver Unit
REP	Right Elliptical Polarization
RCUH	RCU Handler
RF	Radio Frequency
RFI	Radio Frequency Interference
RIME	Radio Interferometric Measurement Equation
RM	Rotation Measure
RMSF	Rotation Measure Spread Function
RO	Radio Observatory
ROB	Royal Observatory of Belgium
RRL	Radio Recombination Line
RS	Remote Station
RSM	Radio Sky Monitor
RSP	Remote Station Processor
SAGECal	Space Alternating Generalised Expectation Maximisation calibration
SAP	Sub-array Pointing
SCO	Station Central Oscillator
SED	Spectral Energy Distribution

SEFD	System Equivalent Flux Density
SERDES	Serializer-Deserializer
SETI	Search for Extraterrestrial Intelligence
SIP	Standard Imaging Pipeline
SKA	Square Kilometre Array
SMBH	Stellar Mass Black Hole
SPAM	Source Peeling and Atmospheric Modeling
SS	Subband Select
SST	Subband Statistics
TAB	Tied Array Beam
TAQL	Table Query Language
TBB	Transient Buffer Board
TDS	Time Distribution System
TEC	Total Electron Content
TECU	TEC Unit
TGSS-ADR1	TIFR GMRT Sky Survey Alternative Data Release
TID	Traveling Ionospheric Disturbance
TIFR	Tata Institute of Fundamental Research
UDP	User Datagram Protocol
UHECR	Ultrahigh energy cosmic ray
VLA	Very Large Array
VLASS	VLA Sky Survey
VLBI	Very Long Baseline Interferometry
VLSSr	Very Large Array Low-frequency Sky Survey redux
VTEC	Vertical TEC
WAN	Wide Area Network
WENSS	Westerbork Northern Sky Survey
WSRT	Westerbork Synthesis Radio Telescope
XST	Crosslet Statistics

Part I
Introduction to Low-Frequency Radio Astronomy and the LOFAR Telescope

Chapter 1
Introduction to Low Frequency Radio Astronomy

George Heald

Abstract Radio astronomy began at low ($\nu \ll 300\,\text{MHz}$) frequencies, but until recently has traditionally been dominated by higher frequency work. With the advent of a new generation of low frequency interferometers and the anticipation of the low-frequency component of the Square Kilometre Array (SKA), the field is experiencing a rebirth. This introductory chapter summarizes some of the key historical developments of low frequency radio astronomy, outlining the science cases that are currently driving the rebirth of the field, and providing an overview of the required instruments and techniques. The role of LOFAR as a key observational facility in the current era of low frequency radio astronomy is highlighted throughout the chapter.

1.1 Historical Context

Radio astronomy is a broad discipline, typically defined to be the branch of astronomy that deals with cosmic radiation observable from Earth at frequencies between the ionospheric cutoff (somewhere around 10 MHz; see Chap. 7) and the tropospheric cutoff in the THz regime. Progressively higher frequencies are part of the infrared portion of the electromagnetic spectrum, as shown in Fig. 1.1. Observations at radio wavelengths can be used to address a broad range of physical phenomena. Broadband continuum radiation is produced by both thermal and relativistic gas: the former through thermal *bremsstrahlung* or "braking radiation"; the latter through synchrotron radiation. Thermal radiation is produced by ionized gas, commonly seen to trace the interstellar medium (ISM) of star forming galaxies. Non-thermal synchrotron radiation is produced by ultra-relativistic charged particles

G. Heald (✉)
ASTRON, Dwingeloo, The Netherlands

CSIRO Astronomy and Space Science, Kensington, WA, Australia
e-mail: george.heald@csiro.au

G. Heald et al. (eds.), *Low Frequency Radio Astronomy and the LOFAR Observatory*, Astrophysics and Space Science Library 426,
https://doi.org/10.1007/978-3-319-23434-2_1

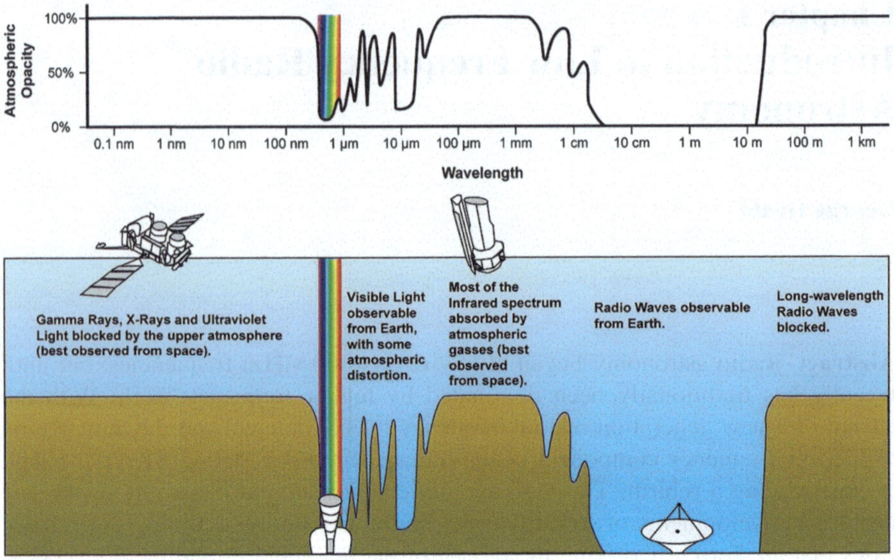

Fig. 1.1 Visualization of the atmospheric transmissivity for electromagnetic radiation. The radio regime spans the range from roughly 10 MHz to 300 GHz, and *low frequency* radio astronomy is defined here to span the range from about 300 MHz down to the ionospheric cutoff around 10 MHz. Image from Wikimedia Commons

(typically electrons) accelerated by magnetic fields, and originates not only in the ISM of galaxies but also in the magnetospheres of pulsars, the intracluster medium (ICM) in galaxy clusters, and active galactic nuclei (AGN). Cyclotron radiation from planetary magnetospheres is observable at the lowest radio frequencies. A particularly useful aspect of radio astronomy is a wealth of spectral lines, from the hyperfine transition of neutral hydrogen (H I) to radio recombination lines (RRLs), molecular transitions and masers.

This book focuses in particular on low frequency radio astronomy. For our purposes, we define "low frequency" radio astronomy as the subdomain dealing with frequencies below about 300 MHz, although the transition from high to low frequency is strongly dependent on context. Radio astronomy is a field that was born conceptually in the domain of theoretical physics as Maxwell and others developed an understanding of electromagnetic radiation, but it was only several decades later when technological developments made observations of the Universe at radio wavelengths possible. In fact the close relationship between technology and the developments in the observational field of radio astronomy is a common theme that we will continue to see throughout this chapter. However, technological developments are both a blessing and a curse to observational radio astronomy: the radio spectrum is also extremely useful for commercial applications such as communications and global positioning. These applications generate radio frequency interference (RFI), a ubiquitous issue that must be addressed in the analysis of data collected by radio telescopes.

1.1.1 Scientific and Technical Background

The so-called "pre-history" of radio astronomy begins with Maxwell's development in the 1860s of a complete theory recognizing electricity and magnetism as manifestations of a single force that also lies at the heart of visible light. Maxwell's work built on decades of progress by Ørsted, Ampère, Faraday, and others. With his equations describing electromagnetism, Maxwell set the stage for a broader understanding of light and electromagnetic radiation more generally. Maxwell proposed the existence of electromagnetic radiation, but it was Heinrich Hertz who first experimentally demonstrated the electromagnetic generation and propagation in free space of radio waves, in the late 1880s. Hertz however did not recognize the full impact of his discovery, stating

> It's of no use whatsoever [...] this is just an experiment that proves Maestro Maxwell was right — we just have these mysterious electromagnetic waves that we cannot see with the naked eye. But they are there.

Other contemporary scientists did recognize the importance of Hertz's discovery: it was not long before radio communications were being developed by engineers like Guglielmo Marconi (the "father of radio") and Karl Ferdinand Braun, co-recipients of the 1909 Nobel Prize in Physics "for their contributions to the development of wireless telegraphy". Interestingly, it was Braun who first developed the concept of a *phased array* (for transmission) in 1905. As we will see, the concept of phased arrays became very important in radio astronomy and they are now a typical feature of low frequency radio telescopes.

The verification of radio waves as a detectable physical phenomenon led to the idea that radio emission from the cosmos might be present and observable. Some early searches were conducted by pioneers including Sir Oliver Lodge, who attempted to detect radio waves from the Sun. These early experiments were unsuccessful, at least in part because they were attempted during solar minimum, with very low sensitivity and in the presence of substantial terrestrial interference—all aspects that were, in retrospect, not ideal. It was through the development of communications technology by Marconi, Braun and others that the prospects of radio astronomy truly became possible. This was particularly true when in the 1920s it was realized that radio communications above about 2 MHz where viable. Before that period, radio communications were focused at frequencies $\lesssim 100$ kHz, and therefore limited by ionospheric opacity. In the 1920s there was a push for receivers operating at tens of MHz, where more voice channels could be accommodated and there was increased reliability thanks to decreased sensitivity to interference and thunderstorms. This shift in communications technology opened the door for concerted research efforts into transmission and reception of radio waves in the regime that we observe today with low frequency radio telescopes like LOFAR. Indeed, as radio communications proliferated, research laboratories were created to develop the needed technology. It is through one of those research facilities, Bell Telephone Laboratories (or Bell Labs for short), that the story of radio astronomy truly begins.

1.1.2 A History of Radio Astronomy

The first recognition of cosmic radio radiation was made by Karl Jansky in 1932. Jansky was working at the Bell Labs site in Holmdel, New Jersey, and was focusing on the identification of noise terms that could be important for a planned transatlantic communication system operating at wavelengths around 10–20 m. Through the use of a rotatable antenna, a Bruce array mounted on tires from a Ford Model T and operating at 20.5 MHz (see Fig. 1.2), Jansky identified three noise terms. Two of those sources of noise were related to thunderstorms at different distances, but the third was a "steady hiss type static of unknown origin" that rose and set daily. Eventually, Jansky noticed that the periodicity of the signal was not diurnal but sidereal, and moreover located approximately in the direction of Sagittarius (e.g., Jansky 1933). He thus concluded that the hiss originated from the Galaxy itself. Radio astronomy had been born. The *New York Times* reported Jansky's discovery on their front page on 5 May 1933, quoting Jansky at the end of the article as commenting that

> There is no indication of any kind [...] that these galactic radio waves constitute [...] some form of intelligence striving for intra-galactic communication.

Given the huge emphasis on communication in the developing field of radio transmission and reception, it would have been natural to immediately consider artificial radio signals as a possible origin of this radiation. But the remark is particularly interesting in the modern era, when researchers working in the field of SETI (the search for extraterrestrial intelligence) are focused in part on the low-frequency radio spectrum as an intriguing region of parameter space (see Sect. 1.2).

Jansky was interested in investigating the Galactic radio emission in more detail, but Bell Labs reassigned him to work on another project. Since the Galactic radiation was so weak, it posed no difficulties in transatlantic communication, and further investment in this area was not deemed to be warranted. Today, a monument to Jansky and his discovery is located at the old Bell Labs site in New Jersey. But it would be a few years before radio astronomy would be pursued further.

One of the many people who were fascinated by Jansky's discovery was Grote Reber. Reber was interested in searching for additional cosmic sources of radio waves, and is responsible for the first ever astronomical sky survey at radio frequencies. During his spare time, Reber built a 10 m parabolic reflector with an elevation axis (a meridian transit telescope; see Reber (2005), where a retrospective has been reproduced) in his backyard in Wheaton, Illinois (see Fig. 1.2). He started his investigations at a wavelength of 9 cm. At the time, the expectation was that the physical origin of the radio waves was thermal blackbody radiation, a hypothesis that Reber's experiments eventually demonstrated to be false. Indeed his early work at 9 cm was unsuccessful, as was his followup at 33 cm wavelength. Reber's persistence paid off when he started observing with a 187 cm receiver (at night, to avoid RFI from automobiles), detecting the Milky Way at the time of year when it was above the horizon during his useful observing sessions. Reber followed up this

Fig. 1.2 The first astronomical radio telescopes. Left: Jansky's telescope, which operated at 20 MHz. Right: Reber's telescope, which operated at several frequencies but successfully detected and mapped the Milky Way at 160 and 480 MHz

early work with sky surveys at 160 and 480 MHz (Reber 1944), confirming that the radiation was nonthermal in nature. Grote Reber was the first radio astronomer, and as the field that he initiated exploded into a great deal of activity in the late 1940s and 1950s, he himself moved on to pursue observations at even lower frequencies, eventually building a dipole array operating at 2 MHz (and with a square kilometer of collecting area!) in Tasmania in the 1960s. Reber was an intriguing personality and the reader is encouraged to read his memoirs on the field of radio astronomy (Reber 1988) as well as about Reber himself (e.g., Kellermann 2004). A full history of the field of radio astronomy is provided by Sullivan (1984).

It is notable that the earliest work in radio astronomy was pursued at low frequencies. As progressively more sensitive receiving systems became available, and radio observations started to become more respectable in the astronomical community, researchers started to move to increasingly higher radio frequencies. This was at least in part because of the need for high angular resolution, which was at the origin of the development of radio interferometry (discussed as a technique in Sect. 1.1.3). Another key development that helped to cement the dominance of GHz frequencies in radio astronomy was the prediction (in 1945 by H. van de Hulst; see Sullivan 1982) and discovery (Ewen and Purcell 1951; Muller and Oort 1951) of the λ21 cm H I spectral line.

Some of the classic radio telescopes that have dominated the field over the past few decades are summarized in Figs. 1.3 and 1.4. From large single-dish telescopes, radio astronomy has largely transitioned to the use of interferometers, but single dishes are still very powerful for certain uses, such as mapping of large-scale diffuse emission (especially over wide areas of the sky) and studies of pulsars and transient sources.

Observations at low radio frequencies have continued despite the historical emphasis on higher frequencies, and the field has enjoyed a resurgence recently, as technological developments have conspired to allow the efficient utilization of

Fig. 1.3 Classic single-dish radio telescopes. Top left: Dwingeloo 25 m telescope (Netherlands ca. 1956, photo © Steve Torchinsky); Top right: Jodrell Bank 76 m Lovell telescope (UK ca. 1957, photo © Cavendish); Middle left: Parkes 64 m telescope (Australia ca. 1961, photo © CSIRO); Middle right: Arecibo 300 m telescope (US ca. 1963, photo © NAIC/NSF); Bottom left: Effelsberg 100 m telescope (Germany ca. 1972, photo © MPIfR); Bottom right: Green Bank Telescope (US ca. 2000, photo © NRAO)

phased arrays of dipoles, as well as sophisticated calibration and imaging techniques needed to overcome, for example, ionospheric effects. Current radio astronomical investigations are able to make use of the full radio spectrum, and the combination of the various frequency ranges is hugely powerful as a scientific tool; see Sect. 1.2. At the lowest frequencies, current efforts are centered around LOFAR (van Haarlem et al. 2013) as well as the Murchison Widefield Array (MWA; Tingay et al. 2013) and the Long Wavelength Array (LWA; Ellingson et al. 2013).

Fig. 1.4 Classic radio interferometric arrays. Top left: Westerbork Synthesis Radio Telescope (WSRT, Netherlands ca. 1970, photo © ASTRON); Top right: Karl G. Jansky Very Large Array (VLA, US ca. 1981, photo © NRAO); Bottom left: Giant Metrewave Radio Telescope (GMRT, India ca. 1995, photo © NCRA); Bottom right: Low Frequency Array (LOFAR, Netherlands ca. 2010, photo © ASTRON)

1.1.3 Review of Radio Interferometry

Here we provide a brief review of the fundamentals of aperture synthesis radio interferometry. Further details can be obtained in standard references such as the "VLA white book" (Taylor et al. 1999) as well as Thompson et al. (2007) and Wilson et al. (2013). Moreover, details specific to the calibration and imaging of low frequency radio interferometers, and LOFAR in particular, can be found in several of the chapters in this book.

The first astronomical use of radio interferometry was actually performed with a *single* antenna placed on a clifftop next to the sea at Dover Heights near Sydney, Australia, and at similar sites in New Zealand. By measuring the interference pattern between the direct path to the antenna and the path reflected from the surface of the sea, Bolton et al. (1949) were able to perform the first identification of radio sources with extragalactic objects (Virgo A, Cygnus A, and Centaurus A). While this was a fascinating technique and provided the earliest indications of the extragalactic origin of the brightest sources of radio emission, the method was not suitable for detailed observations. Instead, the technique of *aperture synthesis interferometry*, based on connected arrays of individual antenna elements, was developed by Martin Ryle and others (see, e.g., Ryle et al. 1950), for which Ryle received the 1974 Nobel Prize in Physics.

Modern radio interferometers are built around a correlator, which combines the signals from the individual antennas making up the array. The signals are delayed relative to each other in order to maximize sensitivity in the direction of the target source (**s**), multiplied and averaged in time. The result of this correlation operation per baseline (**b**, corresponding to a particular pair of antennas) is the visibility V. The visibility is related to the apparent brightness distribution on the sky $I'_\nu(\mathbf{s})$ at frequency ν by the relationship

$$V(\mathbf{b}) = \int\int I'_\nu(\mathbf{s})\, e^{-2\pi i \nu \mathbf{b}\cdot\mathbf{s}/c}\, d\Omega, \tag{1.1}$$

where the integral is over the entire sky. The exponential argument represents an interference "fringe pattern" that is cast onto the sky and multiplies the apparent brightness distribution. The fringe pattern is closely spaced if the antennas are spaced far from each other (i.e. **b** is a long baseline), or widely spaced if the baseline is short. The angle of the fringe pattern on the sky is related to the orientation of the baseline from the viewpoint of the target source. In this way, baselines of different lengths and orientations provide sensitivity to emission on particular angular scales. The rotation of the Earth induces a time variability in the projected baseline length and orientation (e.g., Thompson 1999, see his equation 2-30). The effective angular resolution Θ_{eff} of the array is related to the maximum baseline length $\mathbf{b}_{\text{max}}$ in the array,

$$\Theta_{\text{eff}} \propto \frac{c/\nu}{\mathbf{b}_{\text{max}}} = \left(\frac{\mathbf{b}_{\text{max}}}{\lambda}\right)^{-1}, \tag{1.2}$$

such that it is inversely proportional to the maximum baseline length expressed in units of the observing wavelength.

Equation (1.1) is a Fourier transform relationship between the visibility and the brightness distribution. Thus, if we seek to recover the latter (i.e. make an image of the sky), we need to invert the equation to isolate I'_ν:

$$\tilde{I}'_\nu = \sum_{\mathbf{b}} V(\mathbf{b})\, e^{+2\pi i \nu \mathbf{b}\cdot\mathbf{s}/c} \tag{1.3}$$

This is referred to as the imaging process. The fact that there are a finite number of baselines in an array ($N_b = N(N-1)/2$ where N_b is the number of baselines and N is the number of antennas) is reflected in the fact that the integral in Eq. (1.1) has been changed to a sum. It also means that not all angular scales are sampled, and thus there is a "sampling function" or dirty beam. This dirty beam can be formally written as

$$D_\nu = \sum_{\mathbf{b}} e^{+2\pi i \nu \mathbf{b}\cdot\mathbf{s}/c} \tag{1.4}$$

and our recovered image $\tilde{I}'_\nu$ is the convolution of the true brightness distribution with the dirty beam,

$$\tilde{I}'_\nu = I'_\nu \star D_\nu. \tag{1.5}$$

The process of deconvolving the dirty image is typically achieved with the CLEAN algorithm (e.g., Cornwell et al. 1999). See also Chap. 10 of this volume.

In practice of course there are several complications that must be dealt with before obtaining a final image. Chief among these is the calibration process that seeks to characterize antenna-based errors and simultaneously build up an excellent model of the sky brightness distribution. These issues are addressed in Chaps. 4–9 of this volume.

1.2 Low Frequency Science

Low frequency radio telescopes such as LOFAR address a broad range of science topics. The goal of this section is to present a general overview of the primary science themes that are being studied with LOFAR. It is notable that other current low frequency telescopes have overlapping but distinct priorities; that is mentioned at the end of the section, along with connections to Square Kilometre Array (SKA) science themes and prospects for the future of low frequency radio astronomy.

The LOFAR science case is largely encapsulated in the six Key Science Projects (KSPs) that have guided the development of the project through its design, construction, and commissioning periods. These science cases are:

- Cosmic magnetism
  The study of magnetic fields throughout the Universe.
- Epoch of Reionization (EoR)
  The search for and statistical characterization of an EoR signal.
- High energy cosmic rays
  Detection and characterization of cosmic ray-induced radio showers.
- Solar and space weather
  Study of the Sun, Solar ejecta, and ionospheric activity.
- Surveys
  Deep surveys of the radio sky.
- Transients
  The search for and characterization of transients and pulsars.

Each of the LOFAR KSPs are now briefly described in turn. These descriptions are by no means complete; please refer to the publications produced by each KSP to get a full description of the broad and diverse science themes that are addressed using LOFAR and other low frequency radio telescopes. Another more lengthy summary of LOFAR's KSPs can be found in the LOFAR overview paper (van Haarlem et al. 2013).

1.2.1 Cosmic Magnetism

Magnetism is an important force in the Universe and provides a substantial fraction of the energy density in various objects. Still, the properties of magnetic fields in various classes of sources and the Universe itself remain rather uncertain. The Magnetism KSP (MKSP) seeks to characterize the magnetic fields in various cosmic objects, including nearby galaxies (e.g., Mulcahy et al. 2014), pulsars (e.g., Noutsos et al. 2015), diffuse Milky Way polarization (e.g., Iacobelli et al. 2013), discrete objects in the Milky Way, and perhaps the intergalactic medium (IGM). The strong observational leverage in this area that is provided by LOFAR comes from at least two key effects. First, the typically steep spectrum of synchrotron radiation means that the emission from the objects of interest is very bright at these frequencies (at least in total intensity). Second, the extremely broad wavelength (λ) coverage (or even more strikingly and most relevant to polarization work, λ^2 coverage; see Chap. 10) permits an extremely high precision estimate of the Faraday rotation measure of any polarized emission detectable at low frequency. The counterbalance is that there is typically a great deal of depolarization at such low frequencies, meaning that on balance the density of extragalactic polarized sources is currently much lower than what we know at higher radio frequency.

1.2.2 Epoch of Reionization

At early times, the Universe transitioned from a fully neutral state to an almost completely ionized state. This era is known as the Epoch of Reionization (EoR) and likely occurred somewhere in the redshift range $z = 6.5$–11.5, when the first sources (stars and galaxies) produced enough ionizing radiation to change the ionization state of the IGM. With low frequency observations, the EoR KSP seeks to determine the relative amount of neutral hydrogen H I emission across a range of redshifts and statistically determine the phase transition from pre- to post-EoR. The frequency of redshifted H I is 190 MHz at $z = 6.5$ and 114 MHz at $z = 11.5$. Thus, LOFAR's High Band Antenna (HBA; see Chap. 2) frequency range is perfectly suited to capture the relevant redshift range. By finding the frequency where the underlying noise in extremely sensitive observations changes, this can be identified as corresponding to the redshift at which the IGM transitioned from predominantly neutral to predominantly ionized. From optical work probing the ionization state of the IGM in the foreground of high-redshift quasars (e.g., Becker et al. 2001), the redshift of the EoR should be at $z \gtrsim 6$ ($\nu \lesssim 203$ MHz). Successful detection of the EoR era with LOFAR will require exquisite control of a number of systematic effects, not least of which are the strong foregrounds (namely, astrophysical objects in the field of view) that are orders of magnitude brighter than the cosmological signal itself (see, e.g., Jelić et al. 2010). Research into the EoR is the driver behind a number of observational projects at complementary observatories like the GMRT and MWA, and is the motivation for purpose-built experiments such as PAPER (Precision Array for Probing the Epoch of Reionization; Parsons et al. 2010).

1.2.3 High Energy Cosmic Rays

The cosmic ray (CR) KSP is primarily concerned with studying the properties of ultrahigh energy cosmic rays (UHECR; 10^{15}–$10^{20.5}$ eV) arriving at the Earth. Their observational approach relies on the fact that UHECR encountering the atmosphere produce a shower of ionized daughter particles, which then produce radio emission. By reconstructing the footprint of this radio shower (and its time evolution), various parameters about the CR event can be gleaned, such as the arrival direction, initial particle energy and mass, and the shower radiation mechanism (e.g., Schellart et al. 2013; Buitink et al. 2014; Corstanje et al. 2015; Nelles et al. 2015). The radio observations are supplemented by particle detectors on the ground. These particle detectors are visible on the superterp.

1.2.4 Solar Physics and Space Weather

The Solar KSP uses low frequency radio observations to study the physics of the Sun and its influence on the Earth's magnetosphere (or "space weather"). LOFAR observations can be used both to probe nonthermal radiation from Solar activity in the form of flares and coronal mass ejections (CMEs). Research into solar activity is focused on detailed study of the plasma processes relevant in the solar corona and the initiation and development of flares and CMEs. In the area of space weather the impact of CMEs on ionospheric activity and the consequences for our modern society (e.g. telecommunications) are of prime interest.

1.2.5 Surveys

The Survey KSP performs deep wide-angle surveys of the radio sky. The Survey plans include a three-tiered "wedding cake" structure, beginning with a shallow all-sky survey (Tier 1, or the LOFAR Two-metre Sky Survey—LoTSS; Shimwell et al. 2017), proceeding to a medium-depth wide-area survey (Tier 2), and topped by an extremely deep narrow field of view survey (Tier 3). The science topics covered by this KSP are rather diverse, but the primary science questions include the identification of high-redshift ($z \gtrsim 6$) radio galaxies (expected to be readily found by following up newly discovered steep-spectrum sources), intracluster magnetic fields through the study of diffuse radio emission in galaxy clusters; study of the star formation processes in the early Universe; and the search for new phenomena in the relatively unexplored low frequency radio window. Additionally, standard astrophysical topics are well addressed by wide area LOFAR surveys, including the study of radio galaxies, AGN, nearby galaxies, cosmological questions such as baryonic acoustic oscillations (BAOs), and individual sources in the Milky Way galaxy. For many of these investigations a great deal of power comes from uniting the low frequency surveys performed by LOFAR with the higher frequency surveys that are planned for the coming years like EMU (Norris et al. 2011), APERTIF (Röttgering et al. 2011), and VLASS (e.g., Myers 2014; Murphy et al. 2015).

1.2.6 Transients

The Transient KSP focuses on objects which cannot be described as steady radio sources. In practical terms, the KSP is interested in two broad classes of objects: "fast" transients including pulsars and Fast Radio Bursts (FRBs); and "slow" transient events such as jet sources on all scales, from AGN to stellar mass black holes (SMBHs). The former class is readily identified in beamformed observations (see Coenen et al. 2014, and Chap. 14 in this volume), whereas the latter class, the slow transients, are better found in the image plane (e.g., Carbone et al. 2016), for example through the Transient KSP's Radio Sky Monitor project.

1.2.7 Other Topics

The topics listed here do not cover the full spectrum of science topics that can be addressed by low frequency radio telescopes, which is why general use observatories (see Sect. 1.3) typically incorporate a portion of open-access time in their schedules. A broader overview of the science that can be performed with LOFAR and other low frequency radio telescopes is beyond the scope of this chapter, but the interested reader would find a great deal of material in the proceedings of the 2014 conference "Advancing Astrophysics with the Square Kilometre Array".

1.3 Low Frequency Radio Facilities

As was noted earlier, there is currently a resurgence of interest in the field of low frequency radio astronomy. This is reflected in the fact that there are now several radio telescope facilities that provide the international community with the means to investigate the science topics outlined in Sect. 1.2. In this section we aim to outline the most prominent facilities and their capabilities, as well as discuss the prospects for large steps forward in the near future.

Table 1.1 Overview of major currently available low frequency radio telescope facilities

Telescope name	Frequency range (MHz)	Station/antenna diameter (m)	Number of stations/antennas
GMRT	50,150,235,327	45	30
LOFAR-LBA	10–90	70–87	51
LOFAR-HBA	110–250	30.8–56.5	75
LWA1	10–88	100	1
MWA	80–300	5	256
VLA	230–470	25	27

The most prominent currently available low frequency radio astronomy facilities are summarized in Table 1.1. The majority of these have been recently constructed (e.g. LOFAR, MWA, LWA) or have been recently upgraded (VLA). A few other recently-built telescopes are purpose-built for individual experiments (e.g. PAPER) and are not compared here as general-use telescope facilities. We also do not include single-dish radio telescopes with the exception of LWA, which will eventually be an interferometric array, and the first station of which (LWA1) is currently operational.

Of the telescopes listed in Table 1.1, two are based on steerable antenna designs (GMRT and VLA). These traditional radio telescopes have been supplemented by a new generation of telescopes that are based on phased array technology. Each LOFAR, LWA, and MWA station (and/or tile) is composed of several individual dipole pairs at fixed orientation. The signals from these dipoles are digitized and combined electronically to form virtual beams. Such phased arrays are more flexible: the digital beam forming can in principle lead to arbitrary combinations of pointing direction and observing frequency; but this flexibility comes at the cost of complexity in the downstream processing. These topics are described for the LOFAR case in this book.

LOFAR holds a unique position in the worldwide suite of radio telescopes. LOFAR is amongst the largest radio telescopes in the world, and by some measures is the largest low frequency radio telescope ever built. Moreover, LOFAR allows observations at the lowest radio frequencies observable from below the Earth's ionosphere. The platform that forms the LOFAR radio telescope provides a unique opportunity to push the envelope in testing strategies for flexible multi-beamed imaging and beamformed surveys, deep high angular resolution radio surveys, and more.

The low frequency radio astronomy community is currently using the fresh momentum gained through the strengths of its current facilities to prepare for still bigger and better telescopes in the coming several years. Many of the more forward-looking science cases introduced in Sect. 1.2 can be refined through the use of our existing telescopes, but fundamental progress will require additional capability. Substantial upgrades are already in the works for GMRT and the VLA, and other existing low frequency radio telescopes are also looking toward upgrade opportunities. An extraordinary step forward in the low frequency regime will come from the Square Kilometre Array (SKA)'s low frequency Phase 1 component (SKA1-LOW), which will be located in the Murchison Radio-astronomy Observatory (MRO) in Western Australia. In terms of sensitivity and frequency coverage, SKA1-LOW represents a massive step forward in comparison to LOFAR. Despite the fact that SKA1-LOW will provide transformational new science capability to the radio astronomy community, LOFAR will nonetheless remain unique both in the Northern Hemisphere and at the global level in its niche areas of observing at the lowest radio frequencies ($\nu < 50\,\mathrm{MHz}$) and highest angular resolution ($\theta \lesssim 4''$).

The remainder of this book deals with topics that are relevant to LOFAR users. It is intended to be a handy reference guide throughout the coming years, during which time LOFAR is expected to play a unique and leading role in the continuing history of low frequency radio astronomy.

Acknowledgements I am indebted to several colleagues who have provided their own collected lecture material that I have used as background. Among those I would particularly like to thank Michael Garrett, Jason Hessels, and Ger de Bruyn.

References

Becker, R.H., Fan, X., White, R.L., et al.: Astron. J. **122**, 2850 (2001)
Bolton, J.G., Stanley, G.J., Slee, O.B.: Nature **164**, 101 (1949)
Buitink, S., Corstanje, A., Enriquez, J.E., et al.: Phys. Rev. D **90**, 082003 (2014)
Carbone, D., van der Horst, A.J., Wijers, R.A.M.J., et al.: Mon. Not. R. Astron. Soc. **459**, 3161 (2016)
Coenen, T., van Leeuwen, J., Hessels, J.W.T., et al.: Astron. Astrophys. **570**, A60 (2014)
Cornwell, T., Braun, R., Briggs, D.S.: In: Taylor, G.B., Carilli, C.L., Perley, R.A. (eds.) Synthesis Imaging in Radio Astronomy II. Astronomical Society of the Pacific Conference Series, vol. 180, p. 151 (1999)
Corstanje, A., Schellart, P., Nelles, A., et al.: Astropart. Phys. **61**, 22 (2015)
Ellingson, S.W., Clarke, T.E., Craig, J., et al.: Astrophys. J. **768**, 136 (2013)
Ewen, H.I., Purcell, E.M.: Nature **168**, 356 (1951)
Iacobelli, M., Haverkorn, M., Orrú, E., et al.: Astron. Astrophys. **558**, A72 (2013)
Jansky, K.G.: Pop. Astron. **41**, 548 (1933)
Jelić, V., Zaroubi, S., Labropoulos, P., et al.: Mon. Not. R. Astron. Soc. **409**, 1647 (2010)
Kellermann, K.I.: Publ. Astron. Soc. Pac. **116**, 703 (2004)
Mulcahy, D.D., Horneffer, A., Beck, R., et al.: Astron. Astrophys. **568**, A74 (2014)
Muller, C.A., Oort, J.H.: Nature **168**, 357 (1951)
Murphy, E.J., Baum, S.A., Brandt, W.N., et al.: American Astronomical Society Meeting Abstracts, vol. 225 #113.03 (2015)
Myers, S.T.: Exascale Radio Astronomy, 10302 (2014)
Nelles, A., Schellart, P., Buitink, S., et al.: Astropart. Phys. **65**, 11 (2015)
Norris, R.P., Hopkins, A.M., Afonso, J., et al.: Publ. Astron. Soc. Aust. **28**, 215 (2011)
Noutsos, A., Sobey, C., Kondratiev, V.I., et al.: Astron. Astrophys. **576**, 62 (2015)
Parsons, A.R., Backer, D.C., Foster, G.S., et al.: Astron. J. **139**, 1468 (2010)
Reber, G.: Astrophys. J. **100**, 279 (1944)
Reber, G.: J. R. Astron. Soc. Can. **82**, 93 (1988)

Reber, G.: In: Kassim, N., Perez, M., Junor, W., Henning, P. (eds.) From Clark Lake to the Long Wavelength Array: Bill Erickson's Radio Science. Astronomical Socicty of the Pacific Conference Series, vol. 345, pp. 16–24 (2005)
Röttgering, H., Afonso, J., Barthel, P., et al.: J. Astrophys. Astron. **32**, 557 (2011)
Ryle, M., Smith, F.G., Elsmore, B.: Mon. Not. R. Astron. Soc. **110**, 508 (1950)
Schellart, P., Nelles, A., Buitink, S., et al.: Astron. Astrophys. **560**, A98 (2013)
Shimwell, T.W., Röttgering, H.J.A., Best, P.N., et al.: Astron. Astrophys. **598**, A104 (2017)
Sullivan, W.T. III: Classics in Radio Astronomy. Reidel, Dordrecht (1982)
Sullivan, W.T.: The Early Years of Radio Astronomy - Reflections Fifty Years After Jansky's Discovery. Cambridge University Press, Cambridge (1984)
Taylor, G.B., Carilli, C.L., Perley, R.A. (eds.): Synthesis Imaging in Radio Astronomy II. Astronomical Society of the Pacific Conference Series, vol. 180 (1999)
Thompson, A.R.: In: Taylor, G.B., Carilli, C.L., Perley, R.A. (eds.) Synthesis Imaging in Radio Astronomy II. Astronomical Society of the Pacific Conference Series, vol. 180, p. 11 (1999)
Thompson, A.R., Moran, J.M., Swenson, G.W.: Interferometry and Synthesis in Radio Astronomy. Wiley, New York (2007)
Tingay, S.J., Goeke, R., Bowman, J.D., et al.: Publ. Astron. Soc. Aust. **30**, 7 (2013)
van Haarlem, M.P., Wise, M.W., Gunst, A.W., et al.: Astron. Astrophys. **556**, A2 (2013)
Wilson, T.L., Rohlfs, K., Hüttemeister, S.: Tools of Radio Astronomy. Springer, Berlin (2013)

Chapter 2
LOFAR Overview

Michiel A. Brentjens and Jan David Mol

Abstract LOFAR is the world's largest radio telescope by many measures. It combines signals from phased array antenna stations instead of large parabolic dishes. Signals are digitized at a very early stage, making it a software-heavy facility. The signals recorded by the LOFAR stations are sent to the central processing facility at the University of Groningen's supercomputer centre. The digital back end, named Cobalt, either correlates these data or creates tied-array beams, and stores them on the CEP4 storage cluster for further processing. After that, data are stored at a long term archive, from which users may download them. This lecture introduces LOFAR's architecture and antenna types, provides a detailed overview of the Cobalt digital back end, and discusses phased array beams to a level that should be sufficient to understand the data reduction procedures discussed elsewhere in this book.

2.1 Introduction

The LOw Frequency ARray (LOFAR, van Haarlem et al. 2013) is a very large radio telescope comprising tens of phased-array antenna stations scattered across north-western Europe, with the core of the array situated near the town of Exloo in the Dutch north-east. LOFAR operates at radio frequencies from 10 to 90 MHz (Low Band Antennas or LBAs) and 110 to 240 MHz (High Band Antennas or HBAs).

LOFAR's antenna signals are digitized at a very early stage, right when they enter a station's signal processing cabinet. Digital signals can easily be copied and reprocessed, which makes it possible to observe multiple targets at the same frequency simultaneously. One can also store the raw antenna signals in memory buffers at the station. They can be analyzed whenever a trigger is received. LOFAR can thus observe a few seconds back in time. This capability is heavily used by the

M. A. Brentjens (✉) · J. D. Mol
ASTRON, Dwingeloo, The Netherlands
e-mail: brentjens@astron.nl; mol@astron.nl

G. Heald et al. (eds.), *Low Frequency Radio Astronomy and the LOFAR Observatory*, Astrophysics and Space Science Library 426,
https://doi.org/10.1007/978-3-319-23434-2_2

very high energy cosmic ray group to observe brief radio flashes from cosmic ray air showers at 5 ns time resolution.

The all-digital nature of the instrument requires a massive ICT infrastructure: every second, tens of gigabytes of data must flow to the digital back end at the Centre for Information Technology at the University of Groningen. The output of the correlator needs to be written to a few petabytes of temporary disc storage, and needs to be preprocessed as soon as possible to create the much smaller intermediate data products that are subsequently sent to a distributed long term archive, from which the researcher may retrieve them. When applying for LOFAR time, one not only applies for time on the sky, but also for compute resources as well as storage in the long term archive.

The possibility to retrieve one's data at *any* stage of processing, combined with the rather loose coupling between most software and firmware components in the signal processing chain, make LOFAR less of a radio telescope per se, and more of a radio telescope toolkit from which one composes the optimal low frequency instrument for the job at hand.

2.2 Antennas

A LOFAR station contains two antenna types: the LBA and the HBA, see Fig. 2.1. The LBAs observe the frequency range from 10 to 90 MHz, while the HBAs observe from 110 to 250 MHz. The FM radio band around 100 MHz is filtered out in both antenna designs.

The LBAs consist of two pairs of antenna wires held up by a PVC post, on top of a rebar ground screen. There is an active balun amplifier on top of the post, right where the wires meet.

Fig. 2.1 Low Band Antenna (left) and High Band Antenna tile (right). The inset on the left shows the LBA amplifier that is cast in resin to keep the mice out. The HBA picture nicely shows the styrofoam support antenna structure and the 15 cm space between HBA tiles to allow for water draining and access to the cover anchors

At 50 MHz the radio wavelength is about 6 m. The effective area of dipole-like antennas scales with the square of the observing wavelength. The wavelength corresponding to 150 MHz—a fairly typical frequency at which many HBA observations are conducted—is approximately 2 m. This implies that if one wants at least the same collecting area as the LBA system, one needs at least 9 times the amount of dipole antennas (also called "elements"). No station has remotely enough compute power to combine those signals into coherent beams. It was therefore decided to arrange the HBA elements in "tiles", each containing 16 antennas in a square grid at separations of 1.25 m. The antennas are encased in 5×5 m styrofoam boxes covered in agricultural plastic foil. The signals of the elements within each tile are combined using analogue electronics before the signal is sent to the processing cabinets for digital sampling and beam forming.

2.3 Observing Bands

Unlike most other radio telescopes, LOFAR does not use a local oscillator system to down convert the radio frequency (RF) signal to some lower, intermediate frequency (IF) signal. Instead, the receiver units (RCUs) at the stations directly sample the RF signals from the antennas. According to the Nyquist theorem, LOFAR would need a 500 MHz sampling frequency to measure a 250 MHz signal without any aliasing. If one, however, only wishes to sample part of the 250 MHz band width, a cheaper, slower sampling rate can be used. The LBAs, for example, are only sensitive to radiation at frequencies below 90 MHz.

Figure 2.2 illustrates the selection of LOFAR's possible observing bands. The primary sampling frequency is 200 MHz, although a 160 MHz sampling clock is present too. A 200 MHz clock allows sampling signals with band widths up to 100 MHz only. Signals above that frequency will be aliased (folded) back onto the 100–0 MHz range. To obtain pure 0–100 MHz data for LBA observations, signals above 90 MHz are filtered out *before* sampling. The LBA data fall in the so-called first Nyquist zone because they are all below half the sampling clock frequency. For the LBAs, there is also an option to filter out signals below 10 MHz, or below 30 MHz to remove strong sources of interference in the short wave band.

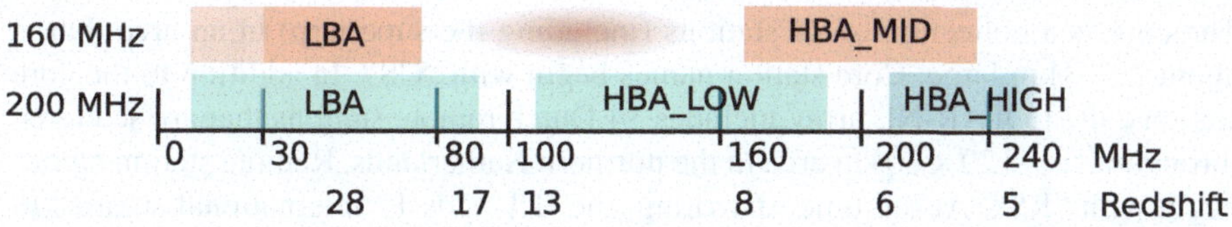

Fig. 2.2 Frequency ranges and clock modes. The second Nyquist zone of the 160 MHz clock is only usable manually and between 110 and 120 MHz, because the 160–200 MHz signals are aliased onto the 160–120 MHz range

For the HBAs, the situation is slightly more complicated. The HBA signals between 100 and 200 MHz are aliased back onto the LBA frequency range in reverse order. They fall in the second Nyquist zone. To prevent pollution by signals below 100 MHz and above 200 MHz, a 110 MHz high pass filter and 190 MHz low pass filter are applied. The 200–250 MHz frequency range is yet again folded back and falls—in correct order—onto the 0–50 MHz frequency range. This is the third Nyquist zone. Here, a 210 MHz high pass filter and 250 MHz low pass filter are applied to reject HBA signals below 200 MHz, and human interference above 250 MHz.

This filter setup, combined with the 200 MHz clock, necessarily excludes the frequency range from 190 to 210 MHz. Enter the 160 MHz clock. The Nyquist zone boundaries of the 160 MHz clock are at multiples of 80 MHz. The 190–210 MHz frequency range is therefore accessible in the third Nyquist zone of the 160 MHz clock. A high pass 170 MHz filter and a low pass 230 MHz filter ensure that signals from other 160 MHz Nyquist zones are rejected.

At the stations, these 100 MHz (or 80 MHz) bands are split into 512 sub bands with band widths of 195,312.5 Hz (or 156,250.0 Hz). When using 16 bits per sample for the beam former and data transport to Groningen, one can use 244 sub bands (48 MHz for the 200 MHz clock). When using 8 bits per sample, this doubles to 488 sub bands (96 MHz for the 200 MHz clock). In principle the hardware also supports a 4 bit mode, however, this mode is not yet commonly usable. It is in principle possible to point every sub band somewhere else. Bandwidth can be traded for beams: one can freely set all sub bands to the same frequency and point them in different directions. A common way to do surveys is therefore to define a set of, for example, 7 beams, using one seventh of the sub bands for each beam, distributing the observed frequencies more or less evenly across the band width available in the chosen frequency range.

2.4 Configuration

As Fig. 2.3 shows, LOFAR is indeed a very large array. For most observation types, LOFAR is used at one of four sizes. From small to large these are the superterp, core, LOFAR-NL, and the International LOFAR Telescope (ILT). The superterp is a cluster of six stations (CS002–CS007) at a circular island that is 320 m across. The core is a collection of 24 stations (including the superterp) in an area that is about 2×3 km large. Core station names begin with "CS". In addition to the core stations, the LOFAR-NL array includes 14 Dutch remote stations that are scattered throughout a $\sim 120 \times 60$ km area in the northern Netherlands. Remote station names begin with "RS". At the time of writing, the ILT adds 13 international stations to the mix: six in Germany (beginning with "DE"), three in Poland (PL610, PL611, and PL612), one in France (FR606), one in Sweden (SE607), one in the United Kingdom (UK608), and one in Ireland (IE613). A fourteenth international station is planned in Latvia. Currently, the maximum ILT baseline is approximately 1980 km.

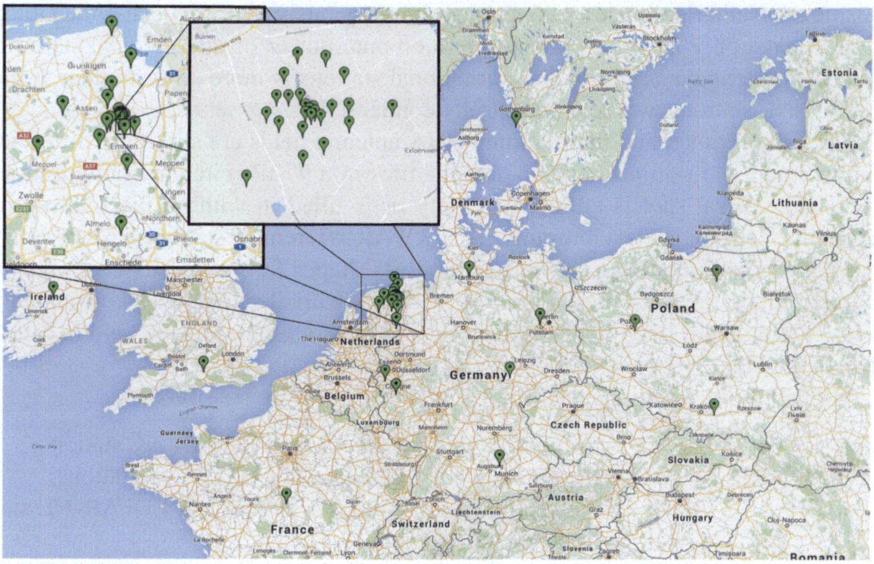

Fig. 2.3 The International LOFAR Telescope has antenna stations throughout north-western Europe. Map data (c) 2015 Google

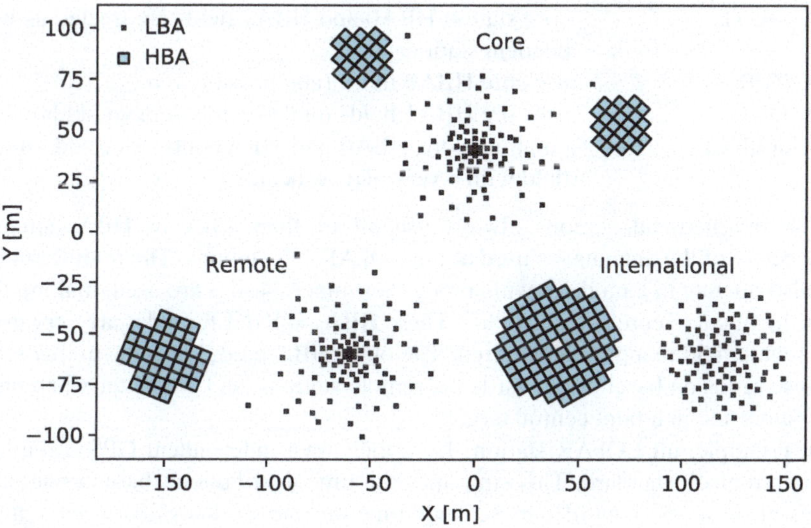

Fig. 2.4 Layout of core, remote, and international stations at the same scale

Figure 2.4 shows the layout of the associated station types. There are 96 LBAs at each LOFAR station. In the Netherlands, only 48 are used at once to form beams, but one can select different sub sets of 48 antennas to suit one's needs. The stations outside the Netherlands use all 96 dipoles simultaneously. This is because the Dutch stations have only half the beam former compute power of international stations. The LBA antenna distribution at international stations is more uniform than at the LOFAR-NL stations, where it is centrally condensed.

The core, remote, and international HBA antenna fields are strikingly different. The international stations have 96 antenna tiles in a 97-tile circular configuration, where the central spot is either left empty or is filled in with an inactive tile to maintain electromagnetic continuity. The LOFAR-NL remote stations have 48 HBA tiles each, approximating a circle by a fat cross shape. To improve uv-coverage on short baselines, the HBA tiles in the core stations are divided into two 24 tile antenna fields, 129 m apart. Each field can be used as a separate station. One field is called HBA0, the other HBA1.

With all those variations, the following antenna sets are possible:

LBA_INNER:	the central 48 LBA antennas at LOFAR-NL stations;
LBA_OUTER:	the outer 48 LBA antennas at LOFAR-NL stations;
LBA_SPARSE_ODD:	the odd-numbered 48 LBA antennas at LOFAR-NL stations;
LBA_SPARSE_EVEN:	the even-numbered 48 LBA antennas at LOFAR-NL stations;
LBA_X:	all X dipoles at LOFAR-NL stations;
LBA_Y:	all Y dipoles at LOFAR-NL stations;
HBA_DUAL:	core station HBA0 and HBA1 fields are treated as independent stations;
HBA_ZERO:	use core HBA0 fields (and possibly remote fields);
HBA_ONE:	use core HBA1 fields (and possibly remote fields);
HBA_JOINED:	combine core HBA0 and HBA1 into one, odd antenna field with a very narrow beam.

The international stations always use all of their LBA or HBA antennas, irrespective of the antenna set used at the LOFAR-NL stations. The remote stations can also operate in a mode in which only their inner 24 tiles are used, making their shape identical to core antenna fields. These HBA_∗_INNER modes are very useful when wide field mapping is required. The other HBA modes provide higher signal to noise ratios on baselines towards the remote stations, and are better when one is only interested in a faint central target.

In principle, all LOFAR stations have their own independent GPS-disciplined Rubidium clock standard. This simplifies the time distribution (there is none), but leaves up to a few tens of nanoseconds time dependent delays between stations. Although this can be corrected post facto in imaging observations using self calibration, it precludes real time tied array beam forming for e.g. pulsar work. Therefore, contrary to the other stations, the core stations all share a common clock signal, distributed to them via glass fibre.

2.5 Signal Path and Data Flow

LOFAR data undergo various stages of analogue and digital processing at various locations (Fig. 2.5). The initial steps of filtering, sampling, sub band separation, and beam forming happen at the stations themselves. The beam formed data are subsequently sent to the central processing facility at the University of Groningen. The digital back end there optionally subdivides the sub bands into channels, and either cross correlates the station streams for imaging, or adds them all together for tied array beam forming.

The data are written to a compute cluster named CEP4, which is responsible for pre processing the data using automated processing pipelines. This involves flagging bad data points, removing the influence of distant bright sources, averaging the data set to a manageable size, and determining and applying calibration solutions from calibrator sources. It is even possible to create rudimentary first-look images there. CEP4 is in principle not freely available to users.

From the central processing facility, data are sent to the long term archive (LTA), from where astronomers can download them to their own facilities for further data reduction. If such a private facility is not available, researchers can request access to LOFAR's guest users facility named CEP3 to reduce their data there.

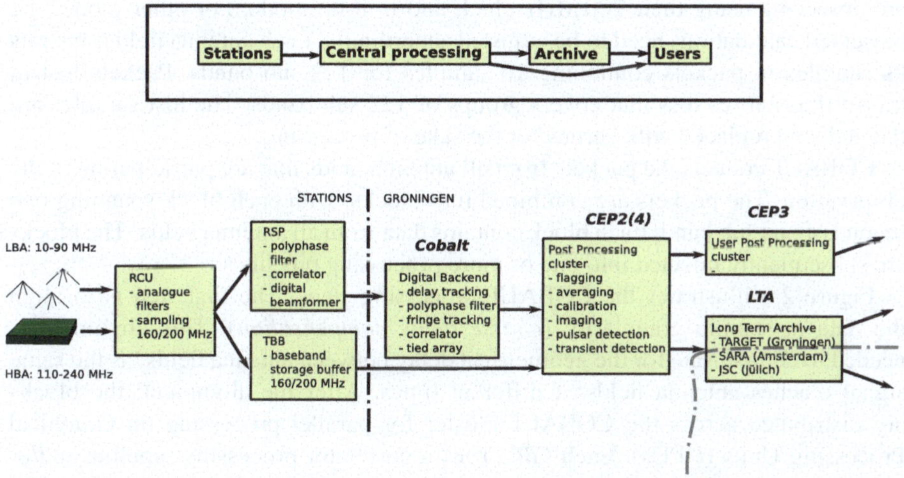

Fig. 2.5 Bird's eye overview of LOFAR data flow. In fact, users can not only obtain their data from the archive and the central processing facilities, but also directly from the stations, although that does involve manual work

More information about central computing facilities is found at ASTRON's radio observatory web site.[1]

It is important to realize that one can actually obtain LOFAR data at *any* stage along the signal processing chain (after analogue filtering, that is), whichever stage is most appropriate.

2.6 COBALT: Central Signal Processing

The LOFAR antenna field's signals are sent to a central processing facility, at which they are combined in real time. The COBALT[2] (Chris Broekema et al. 2018) cluster and COBALT software perform the required signal processing, and are able to correlate the antenna fields and/or form tied-array beams (TABs). This section describes the data flow through COBALT and the processing pipelines it supports. Since COBALT is a real-time system, it needs to keep up with the data stream or suffer data loss. Although data loss is rare for typical observations, it is not something that can always be avoided and will cause specific patterns in the data, which are mentioned in this section.

The signal from each antenna field arrives at a specific node in the COBALT cluster. For this section, we assume for the sake of simplicity that the stations are observing using their 200 MHz clock and in 8-bit mode. For other modes the presented calculations need to be adjusted accordingly. Each antenna field transmits its samples in packets containing 16 samples for 122 sub bands. Packets lost in transit thus causes loss that covers groups of 122 sub bands. The lost samples are flagged, and replaced with zeroes for the sake of processing.

COBALT collects the packets from all antenna fields that are participating in the observation. The packets are combined into blocks, with each block spanning one second of one sub band. Each block contains data from all antenna fields. The blocks are subsequently injected into one or more processing pipelines.

Figure 2.6 illustrates the COBALT processing steps. The first step is to align the data at sample granularity (*coarse delay compensation*). This alignment is needed to compensate for the geometrical delay between antenna fields, as the same signal reaches antenna fields at different times. After the alignment, the blocks are distributed across the COBALT cluster for parallel processing on Graphical Processing Units (GPUs). Each GPU runs a correlator processing pipeline and/or a beam former processing pipeline, according to the observation specification. The pipeline output is then sent back to the CPU for minor post processing.

[1] http://astron.nl/radio-observatory/observing-capabilities/depth-technical-information/cep-and-lta-computing-facilitie.

[2] COrrelator and Beam former Application for the LOFAR Telescope.

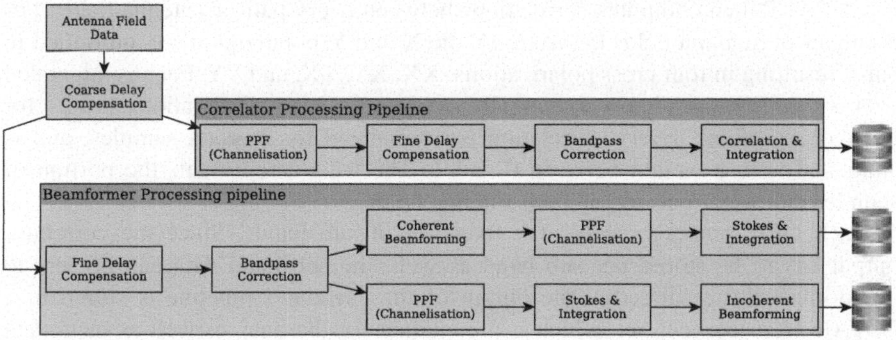

Fig. 2.6 The COBALT signal processing pipelines. Data arriving from the antenna fields is processed by COBALT and sent to disks at CEP4

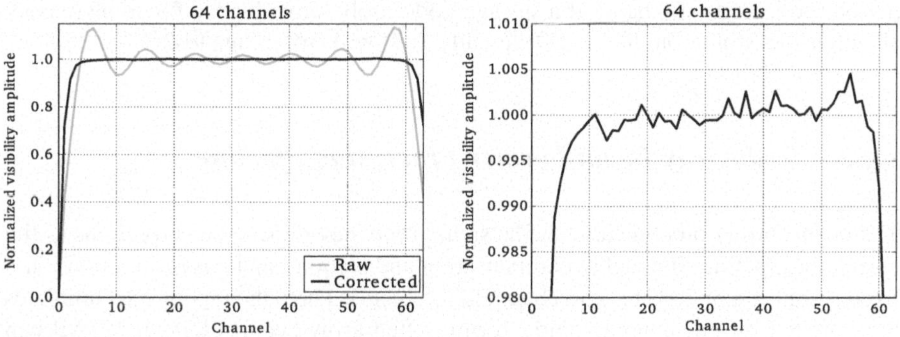

Fig. 2.7 The raw and the corrected sensitivity per sub band, when considering 64 channels/sub band (left), and the remaining error (right)

2.6.1 Correlator Processing Pipeline

The correlator pipeline first obtains the desired frequency resolution by dividing each 195-kHz sub band into *channels* of typically 3 kHz using a poly phase filter (PPF), to cleanly separate the channels within each sub band. Then, for each channel, COBALT applies two frequency-dependent corrections:

1. The sub-sample remainder ($<2.56\,\mu$s) of the geometrical delay is corrected for (*fine delay compensation*), using a frequency-dependent phase shift.
2. The station FIR filter causes fluctuations in sensitivity in each 195-kHz sub band. The *station-bandpass correction* normalizes the amplitude of each channel. Figure 2.7 shows the sensitivity across each sub band before and after correction, as well as the remaining error. Although most of the sensitivity is restored with an error of $<\frac{1}{2}\%$, the sensitivity in the first and last channels remains poor. For that reason, the first and last $\frac{1}{32}$nd ($\approx 3\%$) of the channels are typically flagged in post processing.

COBALT then computes correlations between every pair of antenna fields. For each pair of antenna fields (a *baseline*), the X and Y polarization are multiplied in pairs, resulting in four cross polarizations: XX, XY, YX, and YY. Each is integrated over a specified time interval, typically over 1 s for LBA observations and 2 s for HBA observations. Every correlation is normalized for missing samples, and is annotated with a weight between 0 and 1. The weight represents the portion of samples that were received by both antenna fields.

Each GPU processes data of a fixed set of sub bands. Since the correlator output has to be stored per sub band as well, the sub band data can be sent to the storage cluster directly. The output of the correlator pipeline is stored in a LOFAR *Measurement Set*, which is a container for the data, as well as meta data such as observation settings, flagging information, and which antennas of each field participated in the observation. Any data loss from COBALT to the storage cluster, for example due to network hiccups, result in the loss of one or more integration period for a single sub band. If a storage node itself is overloaded for some reason, all sub bands stored on that node (typically 1–5) will suffer data loss.

2.6.2 Tied Array Beam Former Processing Pipeline

The beam former pipeline applies the same frequency-dependent corrections as the correlator pipeline (fine delay compensation and station band pass correction), at a fixed resolution of 763 Hz (256 channels/sub band). Then, the pipeline accumulates the signals from the antenna fields, forming tied-array beams (TABs). A TAB can be a *coherent*, or an *incoherent* sum of the signal streams. Each coherent TAB has its antenna-field signals aligned to focus on a specific point in the station beam (the antenna field's field of view), and is characterized by a focal direction, typically an RA/DEC J2000 coordinate. For each coherent TAB, COBALT can compute either the raw complex voltages, which is the sum of the antenna field signals, or transform those voltages into either the Stokes I (amplitude), or Stokes I, Q, U, and V parameters.[3] The resulting time series are typically integrated temporally to reduce the data rate. At full resolution, a TAB produces 5.5 TB/h of data. If the TABs are integrated sufficiently, COBALT is capable of computing and storing tens to hundreds of coherent TABs simultaneously.

The incoherent TAB is the sum of the Stokes parameters of each individual antenna field. The incoherent TAB always shares the focal point and field of view of the station beam, but is less sensitive in a specific direction when compared to a coherent TAB. COBALT can output either the Stokes I, or the Stokes I, Q, U, and V parameters of the incoherent TAB, and can apply temporal integration to the resulting time series.

[3]COBALT computes $I = |X|^2 + |Y|^2$, $Q = |X|^2 - |Y|^2$, $U = 2Re(X\bar{Y})$, $V = 2Im(X\bar{Y})$. Note that V is a negation of its formal definition.

The frequency resolution of the TABs can optionally be increased, which is done using a PPF. Since the beam former outputs time series, the increase in frequency resolution directly results in the same decrease in time resolution.

On the storage cluster, each TAB is stored in one or more files, which contain data from all sub bands in the observation. Each TAB is stored in one or more *HDF5 files*, describing the observational settings, as well as the TAB data. The output of each GPU consists of all TABs created per sub band however, so the data need to be reshuffled towards the storage cluster. For that reason, in case of network hiccups, data loss in beam formed observations typically results in flagged samples (zeroes) for one or more sub bands in bursts of $\sim$1 s blocks, with other sub bands for the same time period being received and written successfully. Alternatively, if a storage node itself is overloaded, zeroes are written for all sub bands for each block ($\approx$1 s).

2.7 Phased Array Beam Forming

LOFAR antenna stations are phased arrays: they are aimed by shifting the signals such that waves coming in from the pointing direction are added exactly in phase, see Fig. 2.8. There, the wave fronts arrive first at the left most antenna, last at the right most antenna. The signal of the left most antenna is artificially delayed by this exact amount before being added to the signal from the right most antenna.

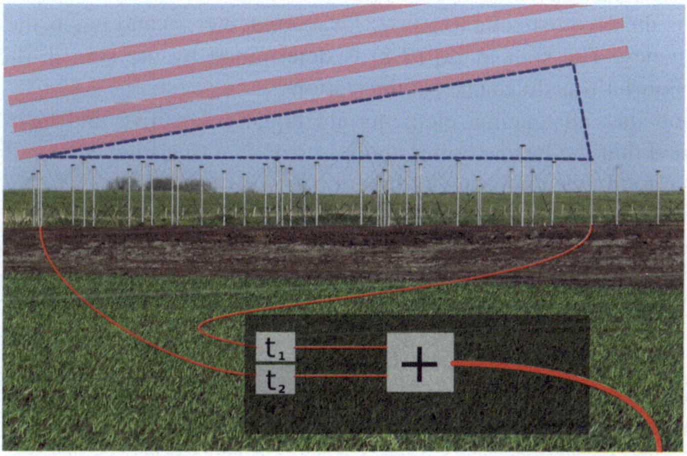

Fig. 2.8 Principle of electronic beam forming

2.7.1 Beam Patterns

The geometric delay between the arrival of a wave front from direction $\hat{\mathbf{l}} = -\hat{\mathbf{k}}$, where $\hat{\mathbf{k}}$ is the unit vector indicating the direction of propagation of the wave, at an antenna at location $\mathbf{x}_n$, relative to arrival at the phase centre $\mathbf{x}_0$ is

$$\tau_n = t_n - t_0 = \frac{(\mathbf{x}_0 - \mathbf{x}_n) \cdot \hat{\mathbf{l}}}{c}, \tag{2.1}$$

where t_n is the time at which the wave arrives at $\mathbf{x}_n$.

Inside the HBA tiles, these delays with respect to the centre of the tile are applied by physically switching lengths of copper into and out of the signal paths of the individual antennas. The digital beam former that combines the signals from the tiles or from the LBA antennas, however does not actually delay the signals. Instead, it multiplies the sub band signals with a complex phase factor

$$\mathrm{e}^{\pm 2\pi \mathrm{i} \nu_m \tau_n} \tag{2.2}$$

before adding the signals. Here, ν_m is the central frequency of sub band m, which is given by

$$\nu_m = \frac{1}{2}\nu_{\mathrm{clk}}(\mathcal{N} + \frac{m}{512}), \tag{2.3}$$

where $\mathcal{N}$ is the Nyquist zone (0 = first, 1 = second, etc...) and ν_{clk} is the sampling clock frequency. The sign in Eq. (2.2) is determined by the sign in the time-to-frequency Fourier transform used at the stations.

Assuming that all antenna elements are equally sensitive and have isotropic response, and that the beam former applies error-free true time delays, the voltage gain in direction $\hat{\mathbf{l}}$ when pointing in direction $\hat{\mathbf{l}}_0$ is given by

$$g_m = \sum_{n=1}^{N} \mathrm{e}^{\mp 2\pi \mathrm{i} \nu_m (\mathbf{x}_0 - \mathbf{x}_n) \cdot (\hat{\mathbf{l}} - \hat{\mathbf{l}}_0)/c}. \tag{2.4}$$

Figure 2.9 shows the LBA and HBA station power beams projected vertically onto the zenithal tangent plane, assuming the beam is pointing straight towards the zenith. These plots take into account the effect of a non-isotropic element beam pattern.

The most striking aspects of these plots are the more or less uniform, noisy structure of the LBA patterns and the regular, symmetric patterns with large areas with low sensitivity in the HBA beams. This is the reason that removal of bright, distant sources is required in all LBA observations, whereas it is only necessary for sources close to the pointing centre in HBA observations.

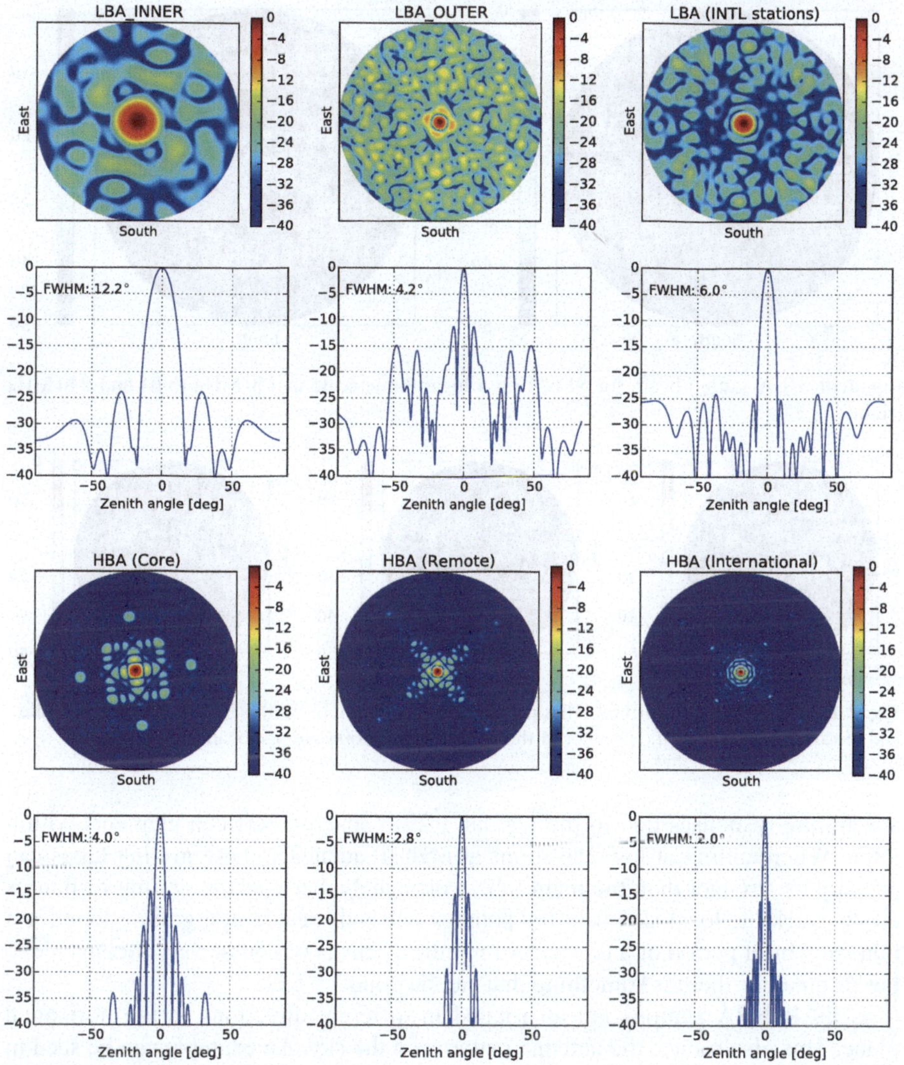

Fig. 2.9 LBA (top) and HBA (bottom) station power beams, which are the square of the voltage beams. Scale is in dB normalized such that the maximum gain is 0 dB. The LBA beams are plotted for 50 MHz, the HBA beams for 150 MHz

2.7.2 Grating Responses

The beam pattern on the sky scales linearly as a function of frequency. Figure 2.10 shows the SE607 HBA beam at 150 and 240 MHz. The −30 dB dots radiating away from the main lobe are grating responses due to the regular 5.15 m spacing between tile centres. The large maxima close to the horizon of the 240 MHz pattern

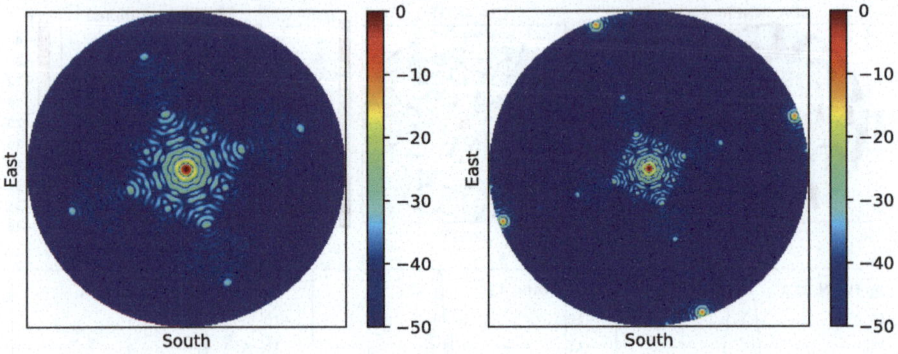

Fig. 2.10 HBA station beam for SE607 on a logarithmic scale at 150 MHz (left) and 240 MHz (right)

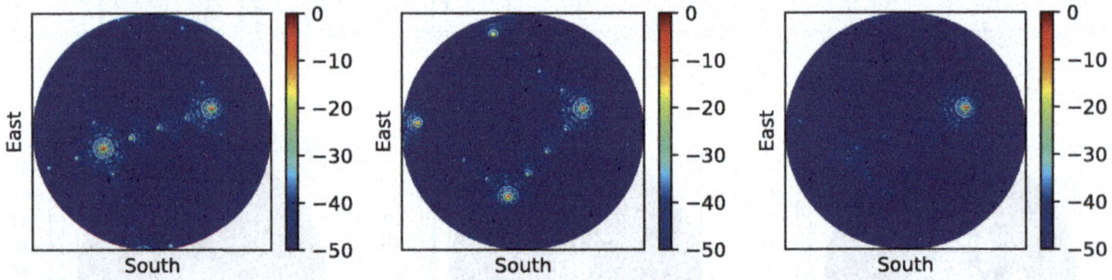

Fig. 2.11 Baseline beam. SE607 HBA power beam at 240 MHz (left), same for UK608 (middle), and the absolute value of the product of the SE607 and UK608 voltage beams (right)

are grating responses due to the regular 1.25 m spacing between elements within a tile. When aiming at low elevations at high frequencies, these grating lobes can actually be stronger than the main lobe. Fortunately, because the grating responses do not track the local sky, celestial sources will only reside in a grating lobe for a relatively brief period of a few tens of minutes. This is of course very inconvenient, but fortunately, there is something that can be done.

LOFAR HBA stations are all rotated in different directions in the horizontal plane. This also rotates the antenna patterns on the sky. An example can be seen in Fig. 2.11. Here, the stations are pointing towards the WNW. For SE607, the grating response is actually stronger than the main lobe. UK608 has a different orientation. Their grating lobes do not overlap. An interferometer baseline between these two stations has a beam pattern that is the product of the voltage beam of one station and the complex conjugate of the voltage beam of the other. The amplitude of that complex valued beam is shown in the right most panel of Fig. 2.11. Because the grating responses do not overlap, they are *vastly* reduced by the low gain at their location in the other beam pattern. The result is a—somewhat counter intuitively perhaps—clean response far from the main lobe.

An effect that does not go away is the reduced sensitivity towards the main lobe that arises whenever a grating response rises above the horizon. To see why this has to happen, it helps to look at the problem as if one is transmitting radio power from the LOFAR antennas. The transmitting and receiving gain patterns are identical. If one transmits radio power at low frequencies, say 120 MHz, no grating responses are above the horizon, so nearly all power is emitted towards the main lobe. At frequencies above ~180 MHz, as soon as a grating lobe is above the horizon, a large part of the power will end up at the grating responses. Keeping the transmitter power constant, this implies that the power that goes towards the grating lobes can only come from one other place: the main lobe, hence the gain towards the main lobe is reduced under these conditions. At the time of writing, this effect is not taken into account in LOFAR's beam model used for self calibration.

2.7.3 A Hierarchy of Beams

The effective station beam pattern can to first order be factorized into the product of several beams, illustrated in Fig. 2.12. For an HBA station there is the beam pattern of an individual dipole, called the "dipole beam" or "element beam". One level up we have the beam pattern of an individual tile, called the "tile beam" or "analogue beam". Finally, these signals are digitally added in the beam former, creating a so-called "digital beam". The station beam is the product of these beams. Note that the "pointing centre" of a dipole is always in the direction of the normal vector of the ground plane. The analogue beam and digital beam can have independent pointing centres wherever one wants to put them. So, in practice, the station beam is the product of these three beams *each shifted to their own pointing centre*.

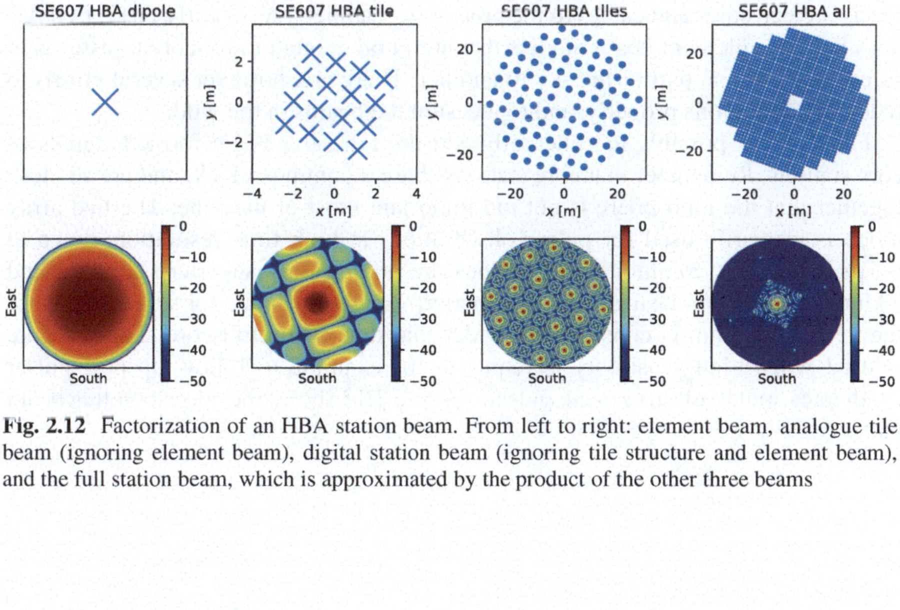

Fig. 2.12 Factorization of an HBA station beam. From left to right: element beam, analogue tile beam (ignoring element beam), digital station beam (ignoring tile structure and element beam), and the full station beam, which is approximated by the product of the other three beams

Observant readers will have noticed that Eq. (2.4) describes a station beam in terms of the discrete Fourier transform of a set of Dirac delta functions at locations $\mathbf{x}_n$. In general, the beam of a beam forming device that records radio waves in some 3D volume, at frequency ν, is

$$g(\mathbf{l}, \mathbf{l}_0) = \int\int\int a(\mathbf{x}) \mathrm{e}^{\mp 2\pi \mathrm{i}\nu(\mathbf{x}-\mathbf{x}_0)\cdot(\mathbf{l}-\mathbf{l}_0)/c} \, \mathrm{d}\mathbf{x}, \tag{2.5}$$

where $a(\mathbf{x})$ is the so-called "complex aperture", which gives the gain and phase with which the incoming signals are distorted as a function of space. If we simply take $a(\mathbf{x})$ to be a set of delta functions at the tile centres, we end up with the beam at the third place from the left in Fig. 2.12. However, we have ignored the internal structure of the tiles. Approximating each tile by a set of 16 delta functions at the locations of each dipole, and assuming that all tiles in a station are equal, *and pointing in the same direction (which is not necessarily the digital beam direction)*, then $a(\mathbf{x})$ is actually the convolution of the set of delta function of one tile, and the set of delta functions describing the locations of the tile centres. Convolution in one domain corresponds to multiplication in the Fourier domain, hence the station beam is better approximated by the product of the digital beam pattern and the analogue beam pattern. Analogously, assuming that all dipoles have the same element beam, one can see that $a(\mathbf{x})$ is in fact the convolution of the set of Dirac delta functions describing the locations of all dipoles, and the complex aperture of an individual dipole antenna, hence the final station beam is the product of the element beam, tile beam, and digital beam.

We know to cm accuracy where all our antennas are, and we know exactly which ones are broken and don't contribute to the beam forming. It is therefore very easy calculate the digital beam and tile beam shapes. Calculating the element beam patterns is unfortunately much more complicated. It involves full electromagnetic simulation of each dipole and its environment, taking into account properties of the receiver electronics and cables to the processing cabinets. As of early 2017, LOFAR does not have element beam models that are good enough to reliably transfer gain solutions from one part of the sky to another. There are, however several efforts to redo old simulations properly, and to measure the beams in the wild.

Finally, it is possible to coherently add the (already) beam formed signals of core stations. Remember that core stations share a common clock, and are so close together that the ionosphere is not too important most of the time. The tied array mode is primarily used for pulsar observations at high time resolution, down to $\sim$5 μs. The most common combinations are adding only superterp stations, and adding the full core. Using only the superterp is a good choice for surveys because the tied array beam is considerably wider that that of the full core. The full core is used when high sensitivity is required, for example to follow up new pulsar candidates, and to observe weak pulsars. Figure 2.13 shows the corresponding beam patterns.

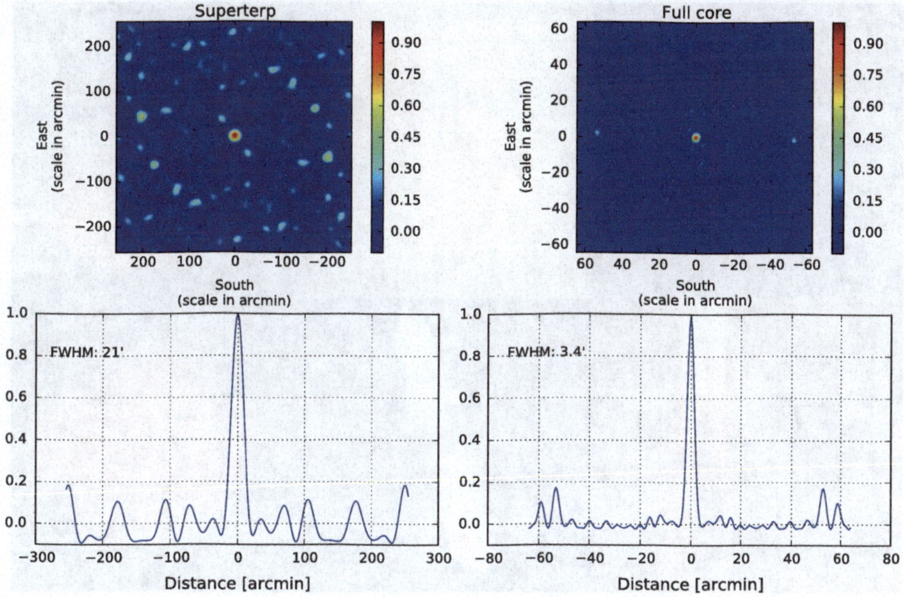

Fig. 2.13 Coherent tied array beam shapes for HBA_DUAL observations at 150 MHz

2.8 Summary

The combination of phased array antenna stations with fully digital processing makes LOFAR a versatile sensor network, from which one can compose the radio telescope one needs. Although phased array beams are quite a bit more complicated than parabolic dish beams, phased arrays are considerably cheaper to construct, and allow one to observe multiple targets simultaneously. LOFAR is unique because raw data can be pre-processed and reduced in volume by automated processing pipelines at the CEP4 cluster, so that astronomers do not have to deal with tens of TBs per night in most cases, and receive an amount that can be stored and processed at a single (heavy) work station. As the remainder of this book shows, LOFAR data reduction is cumbersome, but reasonably well understood, and experienced LOFAR users are now regularly producing magnificent images, such as those in Fig. 2.14.

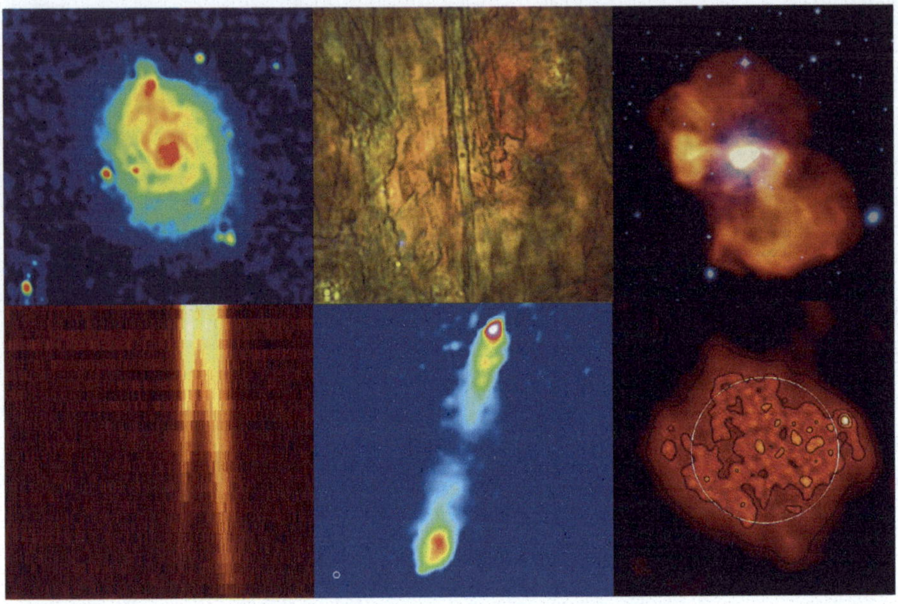

Fig. 2.14 Various recent images and other results made by various LOFAR users. Top row left to right: Messier 51 (Mulcahy et al. 2014), polarized Galactic synchrotron foreground (Jelic et al. in prep.), Virgo A (de Gasperin et al. 2012). Bottom row left to right: Pulsar B0809+74 pulse profile 15–62 MHz (Kondratiev and LOFAR Pulsar Working Group 2013), radio galaxy 3C 223 (Morganti et al., presentation 2014-10-01), the Sun (Mann, observed 2011)

References

Chris Broekema, P., Jan David Mol, J., Nijboer, R., van Amesfoort, A.S., Brentjens, M.A., Marcel Loose, G., Klijn, W.F.A., Romein, J.W.: Cobalt: a GPU-based correlator and beamformer for LOFAR. Astron. Comput. **23**, 180–192 (2018)

de Gasperin, F., Orrú, E., Murgia, M., et al.: Astron. Astrophys. **547**, A56 (2012)

Kondratiev, V., LOFAR Pulsar Working Group: In: van Leeuwen, J. (ed.) IAU Symposium, vol. 291, pp. 317–320 (2013)

Mulcahy, D.D., Horneffer, A., Beck, R., et al.: Astron. Astrophys. **568**, A74 (2014)

van Haarlem, M.P., Wise, M.W., Gunst, A.W., et al.: Astron. Astrophys. **556**, A2 (2013)

Chapter 3
LOFAR Station Processing

Menne J. Norden

Abstract Station processing describes the necessary hardware and software signal processing steps to go from the incoming low frequency electromagnetic signals to stable and good quality scientific output products. The whole analogue and digital LOFAR station processing steps, from the antennas up to the optical fiber output signal, are explained in detail in this chapter.

3.1 Antennas

The antennas are at the beginning of the LOFAR signal chain and they are the key elements which convert the electromagnetic (EM) field into electrical signals. Next, these signals are processed by the rest of the LOFAR system. The sensitivity of the LOFAR system is determined by the number of antennas used to receive the radiation in a given frequency band and by the performance of the antenna elements themselves. Since LOFAR is able to receive radiation in a broad frequency range, there are two types of antenna elements utilised to cover the whole range.

LOFAR was designed and built to explore the low frequency range from 10 up to 240 MHz. To cover this range two different type of antennas had to be designed. One type is optimized to receive signals below the FM radio frequency band (87.5–108.0 MHz) and one for the range above. The antennas are wideband and equipped with Low Noise Amplifiers (LNAs) to amplify the weak astronomical signals. Both antennas are skynoise limited (Norden and Kant 2007). This means that the noise at the output of the system (and consequently at the output of the antenna) has to be dominated by the sky noise contribution. Since the sky noise term increases at longer wavelengths, it allows for a small and cost-effective design. Being sky noise limited implies that the dominant noise source is located outside the system. The

M. J. Norden (✉)
ASTRON, Dwingeloo, The Netherlands
e-mail: norden@astron.nl

G. Heald et al. (eds.), *Low Frequency Radio Astronomy and the LOFAR Observatory*, Astrophysics and Space Science Library 426,
https://doi.org/10.1007/978-3-319-23434-2_3

Low Band Antenna (LBA) contributes 20% of the skynoise whereas the High Band Antenna (HBA) noise contribution equals the skynoise.

3.1.1 LBA

The LBA is extremely wideband. The frequency range covers three octaves (10–20–40–80 MHz). The antenna impedance varies from capacitive (short dipole approximation) to inductive (where the dipole is long compared to the wavelength) and is resonant around 58 MHz. This makes it impossible to impedance match the LNA with the antenna. A compromise is found by using a high impedance single ended amplifier. The amplified signal from the dipoles is combined by an RF transformer.

The LBA features two polarisations. We call them X and Y. Each polarisation is connected to a Receiver unit (RCU). One RCU delivers DC-bias for one polarisation and receives the signal of a single dipole. Two RCUs are needed per antenna, as depicted in Fig. 3.1.

The LBA is placed on top of a 15 cm wire mesh ground plane of 3 m by 3 m. The ground plane acts as reflector and reduces the effect of soil conductivity fluctuations. The droopy dipole (or inverted vee) antenna is optimized for maximum sensitivity to electro-magnetic fields from within zenith angles of ±60°. The dipole length is about 2.4 m and the wires make a 45° angle with the vertical plastic center pole.

The orientation of the X dipole is northeast (NE) to southwest (SW) and the Y dipole is oriented northwest (NW) to southeast (SE), see Fig. 3.2. Each station has a different LBA array rotation so that the station beam sidelobe pattern is misaligned station-to-station across LOFAR. The individual LBA are rotated back to have all X and Y dipoles oriented to the same alignment as seen from an astronomical source.

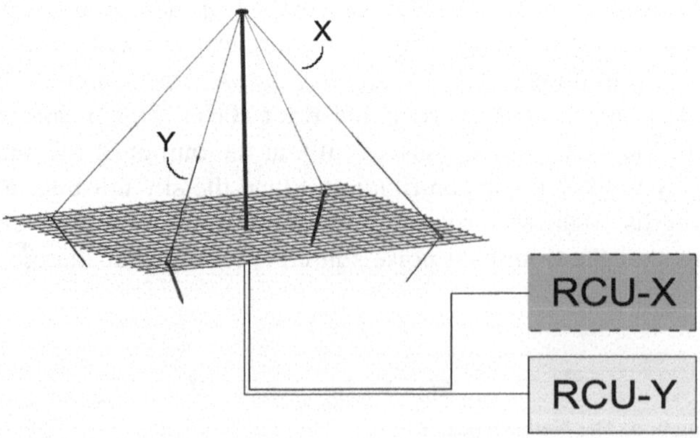

Fig. 3.1 LBA connection

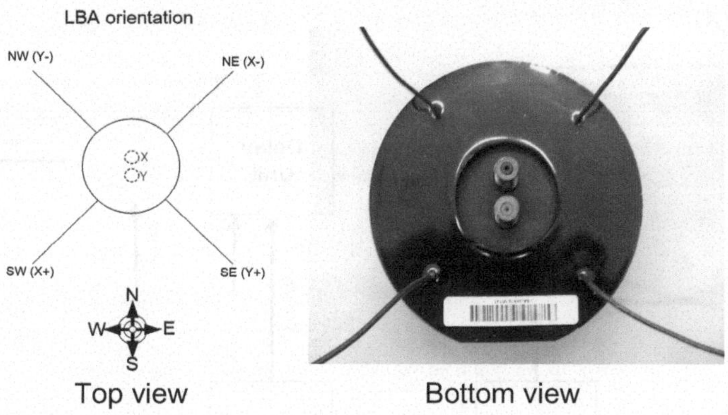

Fig. 3.2 LBA polarity

3.1.2 HBA

The HBA is also a wideband antenna. The frequency range covers a little bit more than one octave (110–240 MHz). The antenna impedance is more constant over the frequency range than in the LBA system. The bowtie dipole element is more wideband by design than the small wire used for the LBA. This makes it easier to impedance match the LNA with the antenna.

On the other hand the noise requirements are more difficult to fulfill, because the skynoise is lower at higher frequencies. To meet the noise and gain requirements a three stage amplifier is used. See Fig. 3.3. After the second stage the signals of the dipoles are combined with an RF transformer. Between the second and third stage a five bit delay line structure is used to direct the pointing of the HBA tile beam.

The five delay lines are placed at the inner layers of the Printed Circuit Board (PCB) of the so called HBA Front End (HBA-FE) unit. The delay unit can be controlled by steps of 0.5, 1, 2, 4 and 8 ns, see Fig. 3.3. The maximum delay is 15.5 ns with a time resolution of 0.5 ns. The delay is sufficient to correct the incoming signal by 4.65 m.

The HBA elements are placed in a regular four by four array. The distance between the elements is 1.25 m. The sixteen antenna elements are placed inside a 5 by 5 m box of Expanded Polystyrene (EPS) foam structures. The EPS structure is held in place by a bottom and top cover of black plastic foil. A 5 by 5 m wire mesh ground plane rests on the inside of the bottom cover. Each side of the HBA Tiles is anchored with two up to four anchors to the ground. The distance between the tiles is 15 cm to allow access for maintenance.

The 16 HBA-FE elements are connected by two 16-1 summators. The power summator (X-Pol) is used to bias all elements inside a tile. The communication summator (Y-Pol) is used to communicate the delay setting to the sixteen HBA-FE's. Each HBA-FE has a unique address (1..16). By adding the "delayed" sixteen elements an analogue tile beam is formed. All tiles in a "mini" array use the same delay setting per antenna element and polarisation (Fig. 3.4).

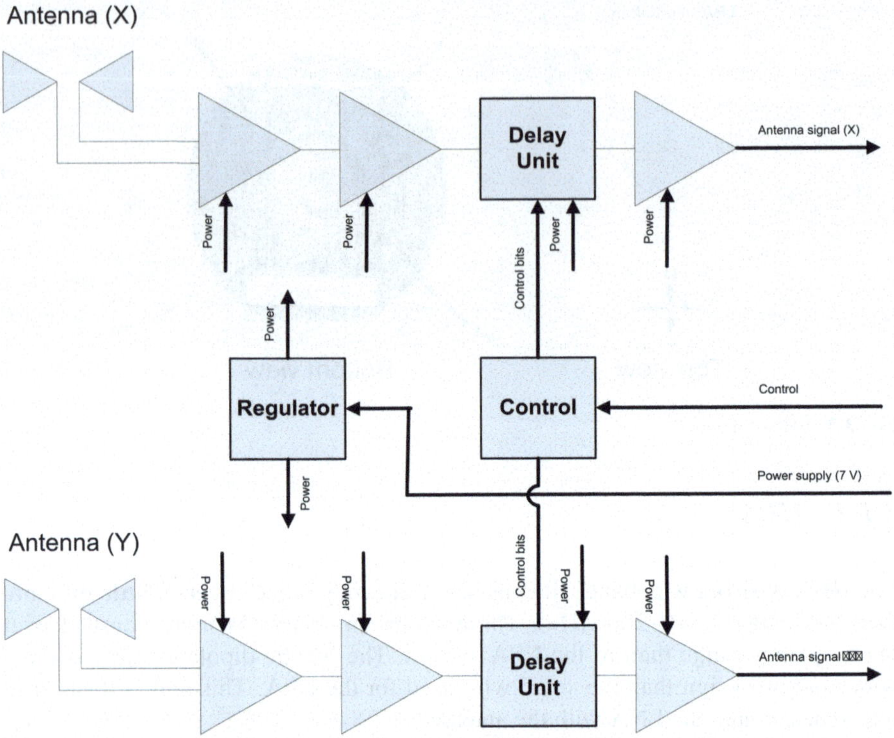

Fig. 3.3 HBA Frontend unit including delay control

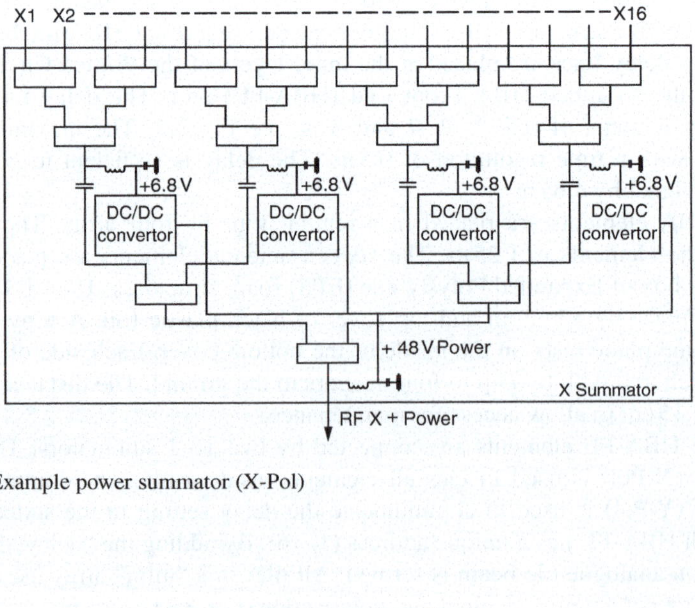

Fig. 3.4 Example power summator (X-Pol)

The orientation of the X and Y dipole is similar to that of the LBA. Each HBA sub-array has a different rotation so that the station beam sidelobe pattern is misaligned station-to-station across LOFAR. As described in Chap. 2, this is particularly important in the HBA because the regular layout induces strong grating lobe responses. The individual HBA elements inside a tile are rotated back so that all X and Y dipoles are oriented in the same direction as seen from an astronomical source.

3.2 Stations

Within LOFAR we distinguish three different types of stations. In the Netherlands we designed two different types (see Chap. 2 for a description of the differences in station layout). The first type is distributed over an area of about 320 ha (3×2 km) close to the village of Exloo. Twenty-four of these so-called CORE stations have been built. Six of them are placed on the central "Superterp" of about 400 m in diameter. These special stations have additional hardware installed. Each station has four LORA detectors that can be used to trigger the Transient Buffer Boards (TBB) for cosmic-ray astronomy (see Chap. 13). An all-sky transient monitoring system is also installed. This Amsterdam-ASTRON Radio Transients Facility and Analysis Center (AARTFAAC) system has its correlator distributed over the six superterp stations. All 288 antennas on the superterp are available for correlation.

The 14 REMOTE station are spread over the northern part of The Netherlands. The longest baseline is 120 km north-south and 60 km east-west. The last type are the international stations. The International Lofar Telecope (ILT) stations are built in Germany (6), Poland (3), France (1), Sweden (1), Ireland (1) and United Kingdom (1). An additional ILT station is planned for Latvia.

Remote and core stations have 2 × 48 LBA antennas, and 48 HBA tiles. An international station has twice the number of antenna inputs compared to a remote or core station. The number of antennas is 96 LBA antennas and 96 HBA tiles. The low frequency band can be observed with different LBA array configurations, each made up of 48 antennas.

The inner array (LBA_INNER) is dense and optimized for the upper frequency range of the low band. The outer array (LBA_OUTER) is sparse and optimized for lower frequency range of the low band. This mode can also be used for observations with smaller field of view. The LBA_SPARSE array is a combination of 24 inner and 24 outer LBA antennas (see Fig. 3.5). Two additional modes, LBA_X and LBA_Y, make use of all 96 antennas but only the X or Y dipole of each, respectively. These may be useful to obtain an excellent station beam, but beware of the effect of differential ionospheric Faraday rotation (see Chap. 7).

On an international and remote station the 96/48 HBA tiles are placed in a regular array. The 48 HBA tiles in a core station are split in two separate regular arrays of

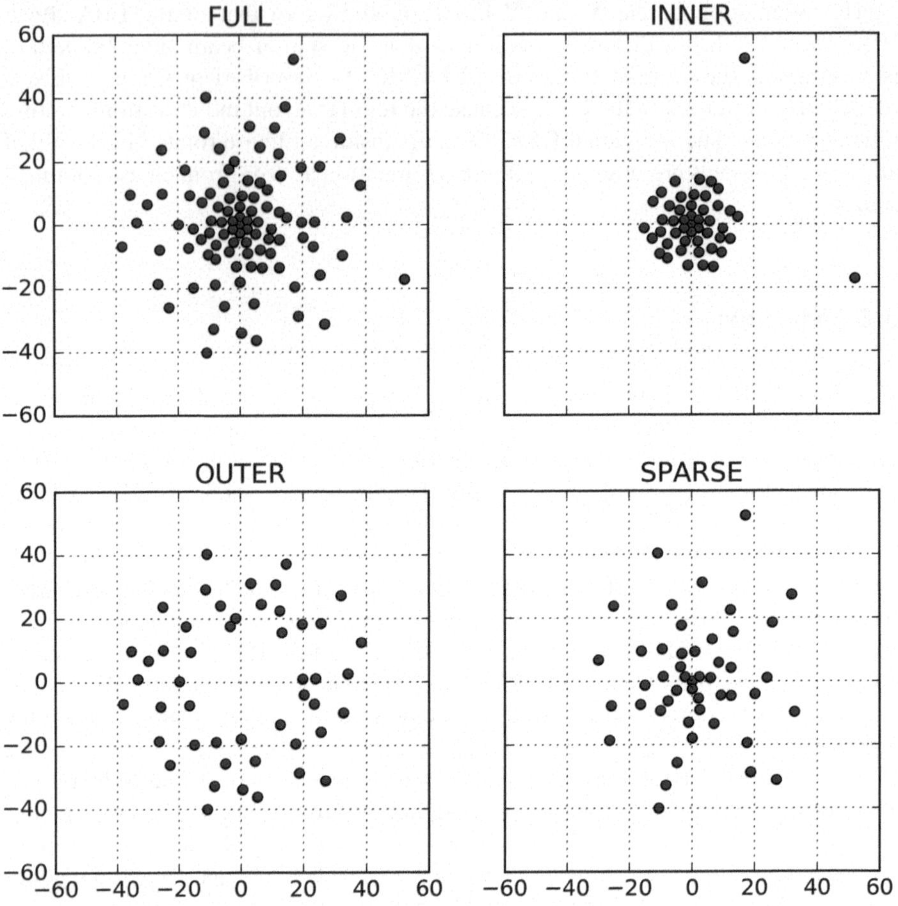

Fig. 3.5 LBA configurations. The axes are local coordinates in meters

24 tiles. A core station has additional hardware (RING SPLITTER) to provide the possibility to split the station into two HBA arrays effectively doubling the number of stations to 48 smaller core stations.

This feature requires double the fiber bandwidth, 6 Gb/s (instead of 3 Gb/s), to the Central Processing (CEP) correlator (see Chap. 2). A common clock signal is distributed by means of a fiber connection to all 24 core stations. A single clock simplifies calibration and improves the coherent addition of station beams in "tied-array" mode because it is not sensitive anymore to clock drift of individual clock systems. Remote and international stations each have individual clock systems. The clock systems will be explained in more detail in Sect. 3.5.

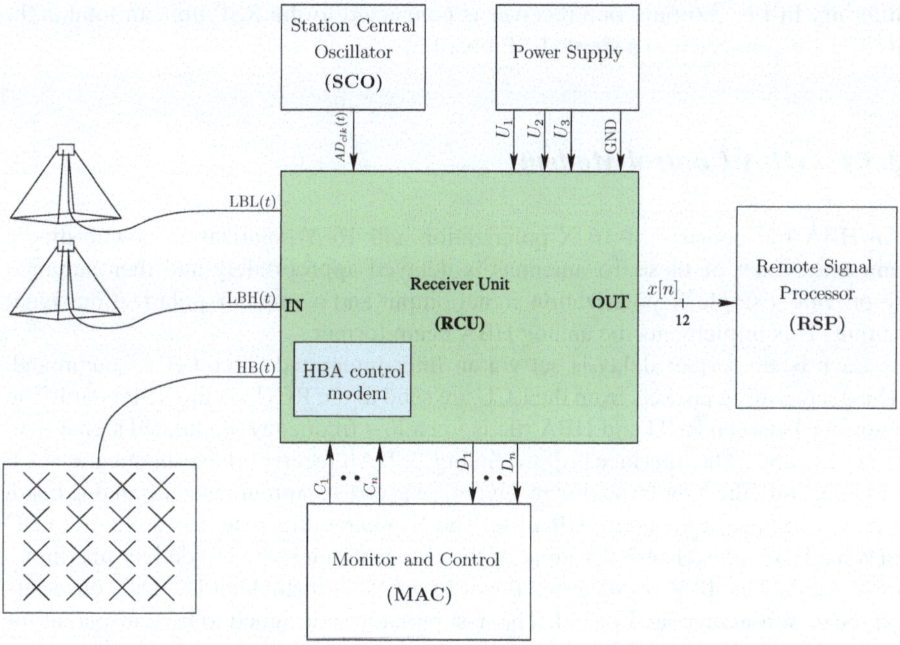

Fig. 3.6 The interfaces of the RCU

3.3 Receiver Unit

The Receiver Unit (RCU) interfaces between the antenna elements and the Remote Station Processor (RSP) board. The function of the RCU is to select one band out of the three input frequency bands (see Chap. 2) and to convert the selected signal towards the digital domain. In Fig. 3.6, the signals are running from left to right. The signals LBL(t), LBH(t) are the LBA signal inputs and HB(t) the HBA signal input. The selected signal $s(t)$ is digitized and represented at the output as $x[n] = x(n \cdot T_s)$, where T_s is the period of the sampling clock $\mathrm{AD_{clk}}(t)$[1] in seconds (s). The sampling clock is provided by the Station Central Oscillator. Nowadays we call the station central oscillator the Time Distribution System (TDS). The TDS delivers the sampling clock for a complete processing unit to the so-called SUBRACK (4 RSP, 2 TBB, 32 RCU).

In order to provide the required functionality and to enable control, additional signals from Monitoring And Control (MAC) and the power supplies are necessary. The digital output signal is passed to the RSP unit for further processing and

[1]In Fig. 3.6, the interior signals in the RCU, like $s(t)$, are not shown. Only the exterior signals, i.e. the signals at the interfaces, are shown.

filtering. In Fig. 3.6 only one receiver is connected to the RSP unit. In total eight RCU's are connected to a single RSP board.

3.3.1 HBA Control Modem

An HBA tile consists of 16 X-polarization and 16 Y-polarization bowtie dipole antennas. Each of these 16 antennas is delayed appropriately and then summed to provide a single X-polarization signal output and a single Y-polarization signal output. This implements the analog HBA beam former.

Each beam former delay is set via an Inter-Integrated Circuit (I^2C) command. The delay setting packets from the LCU are send to the RCU via the RSP board. The interface between RCU and HBA tile is via a low frequency modulated signal over the coax cable. The interface is drawn in Fig. 3.7. All receivers have an integrated LF modem. Only the Y-polarization receivers are used to communicate the analog beam former settings for the entire HBA tile. The X-polarization receivers are used to DC bias the HBA tiles. The HBA input of the receiver is designed to deliver maximum 48 V/1.2 A. The 48 V is converted down to 6.8 V with shielded DC/DC converters on the X-summator, see Fig. 3.4. The Y-summator is designed to be transparent for the communication and RF signals.

The modem communication uses Manchester encoding (used to encode both the clock and the data stream together) with a bit rate of about 5 kbps. The modulation and demodulation function is implemented in the firmware of a microcontroller. For transmit it uses the Pulse Width Modulator (PWM) output and for receive it uses an internal comparator.

To reduce electromagnetic interference (EMI) during observation, both modems can be placed in sleep mode. The modem in the RCU can be woken up by I^2C access and the HBA modem can be woken up by the presence of a preamble. A more detailed description of the HBA control design and communication protocol can be found in Kooistra (2010).

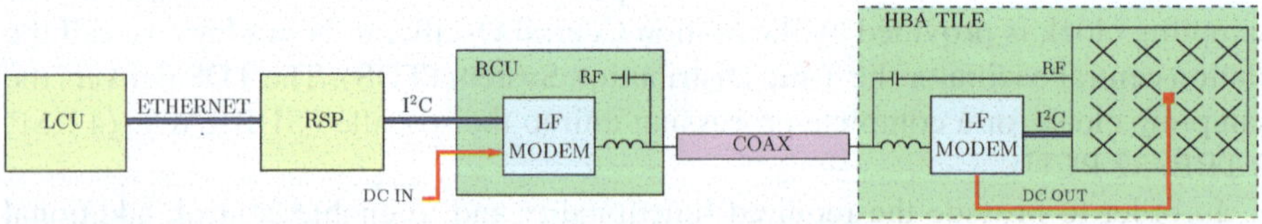

Fig. 3.7 HBA control interface

3.3.2 RCU Modes

A receiver unit has three antenna connector inputs. Not only the input but also the filter type need to be selected. To simplify the selection seven RCUMODEs have been defined. For example in the low band you can select the LBL (LBA_OUTER) or LBH (LBA_INNER) input. An optional 30 MHz high pass filter can be selected to suppress strong RFI in the lower part of the spectrum.

In the high band mode you have the possibility to select three different higher order bandpass filters. In RCUMODE=6 the sample clock also needs to change to 160 MHz. This mode was designed to fill up the frequency gap that would arise otherwise around 200 MHz. When the LOFAR station is not used RCUMODE=0 is selected to save power in standby mode. The different RCUMODES are tabulated in Table 3.1.

The LOFAR receiver does not use a local oscillator or mixer circuit to convert the incoming frequency band. The conversion to baseband (0..100 MHz) is done by a 12 bits Analog to Digital Converter (ADC). To prevent unwanted aliasing the incoming frequency band is bandpass limited by a higher order analog lumped element filter. An analog lumped element filter is a electronic filter with discrete components like capacitors, inductors and resistors (the elements of the circuit). In this way for example the 200–300 MHz frequency band is converted to baseband. The frequency band in rcumode 5 is inverted due to the sub-sampling architecture. This can be corrected in the RSP firmware by using the spectral inversion command.

As an example illustrating how the amplifier and filter chain is constructed a block diagram is given in Fig. 3.8. Each building block is characterized with Gain (G) or Insertion Loss (IL) and Noise Figure (NF). The gain of the two amplifiers (4) and (6) is fixed at 20 dB. The total required gain can be set by defining the attenuation for each PAD. A Π PAD is an attenuator circuit constructed with three resistors in (symbol Π) shape. The attenuation can be set in dB (for example 6 dB). In this way the optimum Noise Figure (NF), isolation between filters and linearity can be designed.

Table 3.1 The 7 RCU modes

RCUMODES	Input	Frequency [MHz]	Clock [MHz]
0	–	–	–
1	LBL	10–90	200
2	LBL	30–90	200
3	LBH	10–90	200
4	LBH	30–90	200
5	HBA	110–190	200
6	HBA	170–230	160 *
7	HBA	210–270	200

* The only RCUMODE with 160 MHz clock signal

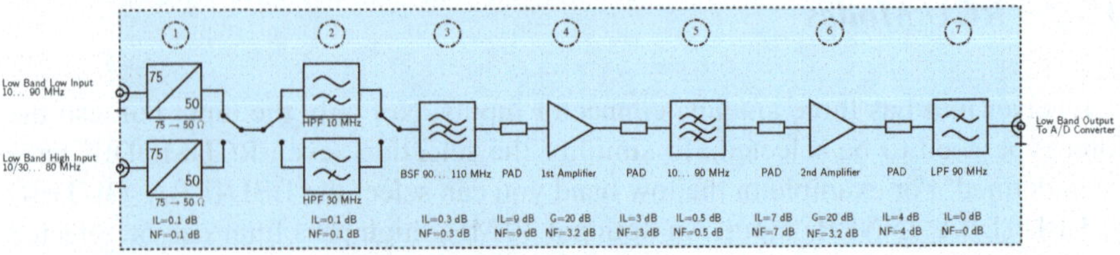

Fig. 3.8 Block diagram of the analog LB receiver chain

The coaxial cables used between the antennas and the receiver have at most three different lengths. This to reduce the length of the cabling for the closest antennas, but results in delay and attenuation differences between the cables. The attenuation differences can be corrected with a 5 bit digital attenuator with steps of 0.25 dB. The delay differences are corrected with steps of 5 ns (200 MHz) in the First In First Out (FIFO) buffers on the RSP board. The remaining phase/amplitude errors are removed by applying a calibration vector for each frequency channel. These calibration vectors are stored in the so-called station calibration tables. For each beamformed observation the calibration table is loaded in order to correct for stable amplitude and phase errors corresponding to each individual dipole signal.

3.4 Digital Processing

The embedded processing of an RSP board filters the digital antenna data, transforms the data into the frequency domain, selects frequency ranges and beamforms them. There is also functionality available to take subband statistics data and to perform station cross correlation (useful for single station imaging, e.g. with AARTFAAC or in single-station mode). The raw digital time samples from the receiver boards can also be sent to Transient Buffer Boards (TBB) to capture and trigger on transient events. An RSP board consists of five Field Programmable Gate Arrays (FPGAs). Four antenna processor (AP in Fig. 3.9) FPGAs which processes the antenna data, and one board processor (BP) FPGA which handles board control, data output and external interfaces. The FPGAs are connected to each other by the internal ring, a high speed LVDS (low voltage differential signalling) parallel interface. RSP boards are connected to each other using the interboard interface which is a high-speed connection using SERDES (serializer-deserializer) components. Two ethernet interfaces for LCU (control) and for CEP (data) are present. Each antenna processor is connected with two receivers giving the full 100 MHz baseband input signal from two antenna polarisations. The setting and readout of the correct receiver mode is done by control interface between the AP and the receivers, see Fig. 3.9.

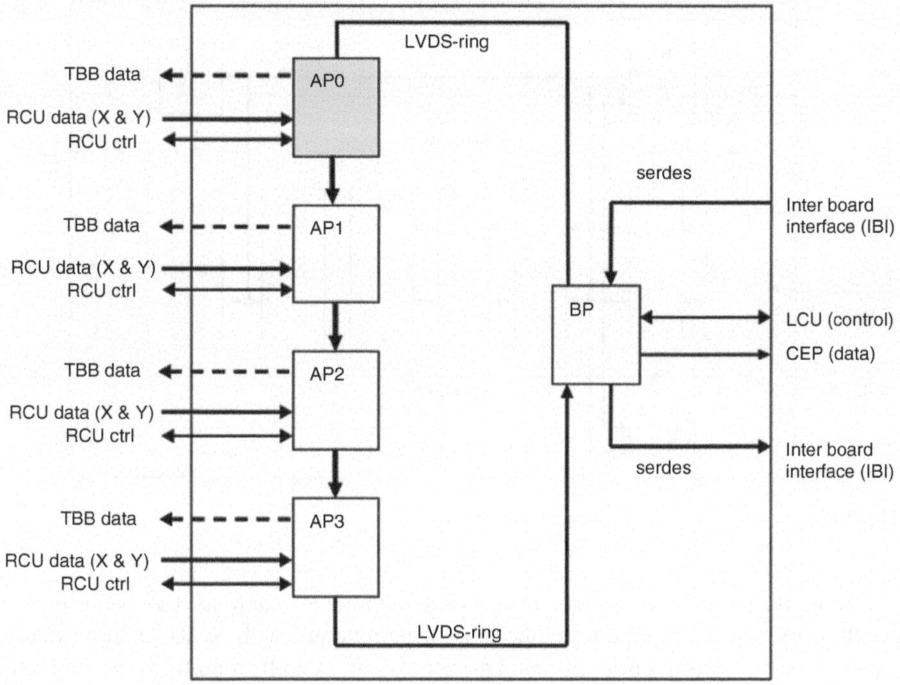

Fig. 3.9 Functional layout of the remote station processing board

3.4.1 *Beamlet Processor (BLP)*

The beamlet processor (see Fig. 3.10) is a module inside the AP that processes RCU antenna data into beamlet data. The data from RCU is received by RCU Handler (RCUH) and then synchronised to the rest of the system by the Block Synchronisation (BS). Small delay differences between the AP's are corrected with 64 steps of 75 ps. For each AP this correction is determined with the so-called PPStune script. The Pre Filter Structure (PFS) performs a Finite Impulse Response (FIR) prefilter operation to shape the data for the transformation into the frequency domain by a Pipelined FFT (PFT). The spectrum data that is required for the measurement is selected by Subband Select (SS) and finally the data of two polarizations is beamformed by BF. Up to the PFT the data is processed in parallel, independently for the X and Y signal. After the PFT the data comes out in parallel too, one real stream with alternating X and Y and one imaginary stream with alternating X and Y. The SS outputs one stream of subbands $\{X_{re}, X_{im}, Y_{re}, Y_{im}\}$. The unit processing block is set by the FFT size of 1024 points and called a slice. This implies that the maximum number of subbands in the SS output stream is 1024/4 = 256. Due to frame overhead in the ring adder (RAD) the actual maximum becomes 244 subbands in 16 bit mode.

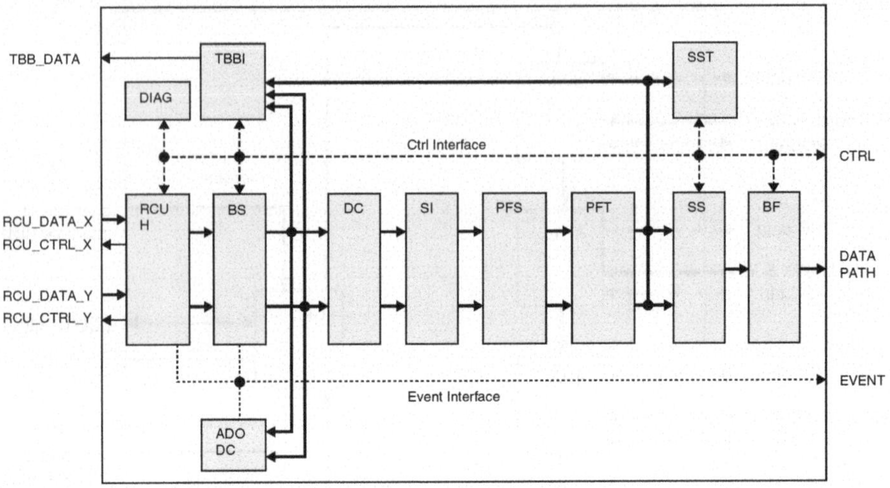

Fig. 3.10 Functional decomposition of the BLP module

Using the control interface, processing parameters such as the beamformer coefficients can be changed, or operational parameters such as RCU input delay can be modified. Frequency spectrum power statistics can be read from the subband statistics (SST) block. The event interface is used for such events as RCU data overflows, or sync pulse synchronisation errors.

3.4.2 Subband Statistics (SST)

The subband statistics contain the real powers for each subband added over 1 s, and updated every second. Note that the SST are available per antenna, so in total 512 subbands × 96 antennas × 2 polarizations. The SST values are real powers because $(a+ib) \times (a-ib) = a^2 + b^2$, where $(a+ib)$ is a subband sample. A subband is defined as a dual polarization complex signal, so $\{X_{re}, X_{im}, Y_{re}, Y_{im}\}$.

3.4.3 Beamlet Statistics (BST)

The beam forming applies weights to each subband and then adds the corresponding subbands from all antennas. The output of the beamformer (BF) is called a beamlet, hence a beamlet is a beam formed subband. A beamlet is defined as a dual polarization complex signal, so $\{X_{re}, X_{im}, Y_{re}, Y_{im}\}$. There are 244 beamlets per station in the 16-bit beam forming mode. At 200 MHz sampling rate the subband and beamlet bandwidth are 200 MHz/1024 = 195,312.5 Hz. This yields a station output beam bandwidth of 244 × 195,312.5 = 47.65 MHz per polarization, which is more than the design requirement of 32 MHz. Nowadays we can also select 4 or 8 bit mode beam forming. This results in 976 or 488 beamlets that can be used to increase bandwidth or total number of beams.

The beamlet statistics contain the real powers for each beamlet added over 1 s, and updated every second. Note that the BST is available per station (per polarization), so in total 244 powers × 2 polarizations.

The beamlet data to the Central Processor (CEP) is distributed over four serial output channels. For each output, a separate 'LANE' is used in the board-to-board SERDES data link. Each lane has a different start RSP board. The end RSP board sets the RSP board that will output beamlet data to CEP. Note that the beam forming operation is also what reduces the data rate from the stations towards CEP.

3.4.4 Crosslet Statistics (XST)

For one subband the station digital signal processing hardware is able to calculate the complex cross powers between antennas, this subband is called a crosslet. For a station with 96 dual polarization antennas this yields an XST matrix of 192 by 192 complex powers. The cross powers are added during one second and updated every second. By selecting another subband as crosslet every second it is possible to get an XST cube of 192 × 192 × 512 cross power values in about 8.5 min (512 s). The data cubes can be used to calculate the calibration vector for each of the 512 subbands. Note that the XST matrix diagonal contains the auto powers, i.e. the same values that are available in the SST of the antennas for that subband. Furthermore half of the XST matrix is the complex conjugate of the other half, so redundant. In other words, the cross correlation power of antenna pair p, q (or $\mathtt{xpow}(p, q)$) equals the power of the complex conjugate of antenna pair q, p ($\mathtt{xpow}(q, p)$). The XST values are complex powers because $(a+ib) \times (c-id) = ac+i(bc - ad) + bd$, where $(a+ib)$ is a subband sample from one antenna and $(c-id)$ is a subband sample from another antenna.

The cross-correlation function is distributed over four serial lanes. On each RSP board, the local crosslets obtained from the APs are cross-correlated on the BP with the remote crosslets, and inserted into the data stream. For each crosslet, a separate slot is allocated in the data frames that are exchanged between the RSP boards. For the crosslets all four lanes need to start on the same RSP board to avoid that some correlations products do not get calculated.

3.5 Station Clock

The remote and international stations are too far away from each other to have a single centrally distributed clock and Pulse Per Second (PPS). Instead at each remote or international station the Station Central Oscillator (SCO) generates a local clock.

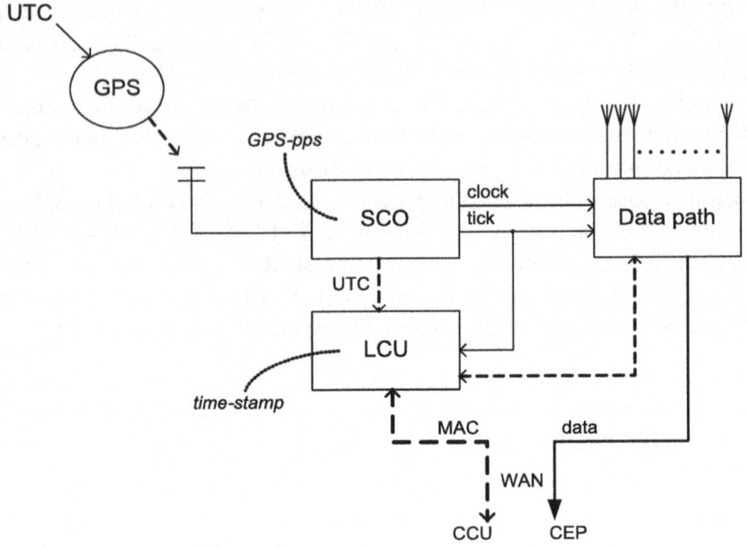

Fig. 3.11 Remote station timing and synchronization overview

The official world time (UTC) information to initialize the pps and to discipline the clock is obtained via GPS satellites. Note that by synchronizing all stations to UTC they are also mutually synchronized. The pps accuracy of a LOFAR GPS receiver is on the order of 10 ns rms. This pps signal is connected to the 1 pps input of a rubidium clock. The rubidium clock itself takes care of the short term clock stability and for the long term it is disciplined by the pps signal from the GPS receiver.

Figure 3.11 shows the SCO in a remote station with GPS-pps as input and clock and tick outputs. The tick in a remote station is 1 pps. The LCU runs a tick counter to provide the time-stamps. The wide area network (WAN) connects the LCU to the LOFAR Central Control Unit (CCU) for MAC communication and it transports the station data output to CEP. For the time-stamping of the data to CEP the residual inaccuracy in the tick needs to be calibrated out at CEP.

The TDS board takes care of generating the 160 or 200 MHz sampling clocks and distributing them into the data path. The 160 or 200 MHz clocks are locked to a 10 MHz reference from the rubidium, and the tick identifies the common clock cycles in the 160 or 200 MHz clock domain. The tick counter on the LCU is started on a rising edge of the pps signal and resets when the corresponding 160/200 Mcycles are counted. During the first few years of operations we noticed that the temperature stability of the rubidium clock, despite using a temperature-controlled chamber, was not sufficient. The pps and 10 MHz output signals drift when temperature changes from day to night, and winter to summer. To prevent this drift, resulting in lower station sensitivity, we installed a Syncoptics repeater between rubidium clock and the TDS boards (Schoonderbeek 2012). The pps and

10 MHz signals are phase locked by the Syncoptics repeater and therefore become temperature independent.

The core stations are located within 3 km distance from the concentrator node. For the 24 core stations we make use of a single centrally distributed clock and pps. For this clock system we installed in the concentrator node a GPS receiver, rubidium clock and LCU. The pps and 10 MHz from the rubidium are connected to the input of a Syncoptics transmitter. The Syncoptics transmitter has 10 optical outputs delivering a combined 10 MHz and pps signal on a 1550 nm optical carrier. Three of these units are needed for all core stations. At each core station a Syncoptics receiver is installed to distribute clock and pps by coaxial cables to the TDS boards. The SCO is completed with a GPS receiver (only connected with the LCU) and a serial cable from the Syncoptics receiver to feed the pps tick signal into the LCU. The single clock, with proper inter station delay calibration, eliminated clock difference between the core stations. Also the signal quality for coherent adding tied-array beams on the core stations is greatly improved.

References

Kooistra, E.: LOFAR ASTRON Memo Series, LOFAR-ASTRON-MEM-175, Rev 2.1 (2010)

Norden, M.J., Kant, G.W.: LOFAR ASTRON System Requirement Specifications, LOFAR-ASTRON-SRS-014, Rev 1.2 (2007)

Schoonderbeek, G.W.: ASTRON Requirements Package, ASTRON-RP-479, Rev 1.3 (2012)

Part II
Fundamentals of LOFAR Data Processing

Chapter 4
RFI Flagging, Demixing and Visibilities Compression

Tammo Jan Dijkema

Abstract The raw correlated visibility data that comes out of the correlator contains radio-frequency interference: signal originating not from celestial sources but, for example, television signals. To enable removing the unwanted interference, the raw data is stored at very high resolution. On this high resolution, the data can be flagged. Also, some more unwanted signal can be removed, namely from very strong celestial sources that are not in the field of view. Demixing is a technique to remove their signal in a relatively cheap way. Once the unwanted signals are removed at high resolution, the data can be averaged down in time and or frequency. All these steps will be discussed in this chapter.

LOFAR data is usually obtained at a high resolution in time and frequency. This chapter discusses the reasons for this, and describes the program DPPP which is used to pre-process and average the correlated data, for more efficient later processing. DPPP was developed by Vishambhar Nath Pandey and Ger van Diepen.

4.1 Best Discretization in Time and Frequency

The integration time for correlation depends on the field of view and the resolution of the image. The integration time Δt should satisfy (see e.g. Condon and Ransom 2007, Thompson et al. 2008)

$$\Delta\theta\,\Delta t \ll \frac{\theta_s P}{2\pi} \tag{4.1}$$

T. J. Dijkema (✉)
ASTRON, Dwingeloo, The Netherlands
e-mail: dijkema@astron.nl

G. Heald et al. (eds.), *Low Frequency Radio Astronomy and the LOFAR Observatory*, Astrophysics and Space Science Library 426,
https://doi.org/10.1007/978-3-319-23434-2_4

where P is the sidereal rotation time of earth (about 24 h), $\Delta\theta$ is the field of view, and θ_s is the size of the synthesized beam. If you want an image of, say, 500 synthesized beams in diameter, an integration time of about 10 s should suffice. Larger integration times will cause a tangentially smeared synthesized beam.

For the discretization in frequency, the channel width $\Delta\nu$ should satisfy

$$\Delta\theta\,\Delta\nu \ll \theta_s\,\nu \tag{4.2}$$

Again, for an image with 500 beams in diameter, and assuming a frequency of around 150 MHz, a channel width of about 150 kHz should suffice. Larger channel widths will cause a radially smeared synthesized beam.

However, there are two reasons that the discretization in both time and frequency used in the correlator is much finer:

- Bright sources outside the field of view corrupt the signal; they have to be removed from the data. To be able to remove them, the signal originating from these sources must not be too much smeared in time and frequency. This means that (4.1) and (4.2) should hold for a field of view where the bright sources are included.
- RFI that occurs within small ranges of time and frequency can be removed at those small scales. In this way, we avoid having to throw away too much data.

After the strong signal and RFI have been removed, the data can be averaged down in frequency and time, to save disk space and speed up further processing. A common scenario is also to start analyses with averaged data to develop a good imaging strategy, and afterwards perform this strategy on higher resolution data.

4.2 Default Preprocessing Pipeline: DPPP

For LOFAR, the above mentioned steps are handled by the program DPPP (Default Preprocessing Pipeline). As the name suggests, this program can handle a pipeline of several steps. It reads and writes the data only once.

The program DPPP is configured by a *parset* (parameter set). This usually is a file, where each line contains a key-value pair. An example parset for DPPP is given by

```
msin=L114220.MS
msout=L114220_processed.MS
steps = []
```

The parset above instructs DPPP to read the file `L114220.MS`, convert it to the standard CASA format, and save the result as `L114220_processed.MS`. Conversion to the standard format is necessary because the LOFAR correlator stores its visibilities in a nonstandard way that is optimized for write speed.

If the above commands are saved in a file, say `myreduction.parset`, the command `DPPP myreduction.parset` will run DPPP with the specified steps. Alternatively, the parset can be given on the command-line.

4.3 Flagging RFI

The radio spectrum, at least in the populated area of LOFAR, is regulated by the government. Only a very small number of frequencies in the radio spectrum are reserved for radio astronomy: no one is allowed to transmit at these frequencies. On the majority of the LOFAR frequencies, the spectrum has to be shared with other services, such as mobile telephony, airplane radar, portophone, semafone, digital radio, and FM. Indeed, the signal transmitted by FM-radio from 87.5 MHz to 108 MHz is so strong that LOFAR cannot observe any astronomical signal there. Luckily, the interfering signals on other frequencies can often be detected and removed from the observation, so that the remaining data is only from the sky.

From Fig. 4.1, it is clear that almost all LOFAR data is to some extent affected by RFI. The good news is that the total amount of RFI usually is below 5%, so there is still more than 95% of unaffected data left. The task we are facing is *flagging* the RFI signal, thus excluding it from further processing.

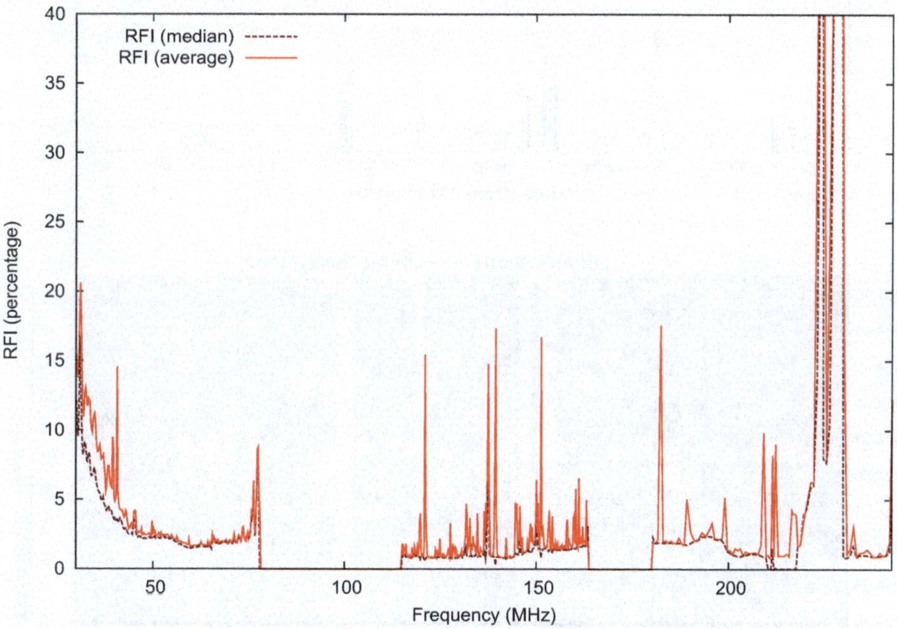

Fig. 4.1 The RFI usually found in LOFAR data, from the LOFAR Imaging Cookbook (Shulevski 2017). The peak in RFI around 220 MHz is caused by digital radio

RFI usually satisfies one or more of the following three properties:

- The signal from interfering sources is orders of magnitude stronger than that the signal of astronomical sources.
- The interfering signal is localized in time.
- The interfering signal is localized in frequency.

4.3.1 Inspecting Data to Spot RFI

To get an idea about how affected by RFI the data is, one could look at a plot of amplitude vs. time of the visibilities. Already this plot reveals RFI that is localized in time, see Fig. 4.2: at certain times, the amplitude is much larger than most of the time.

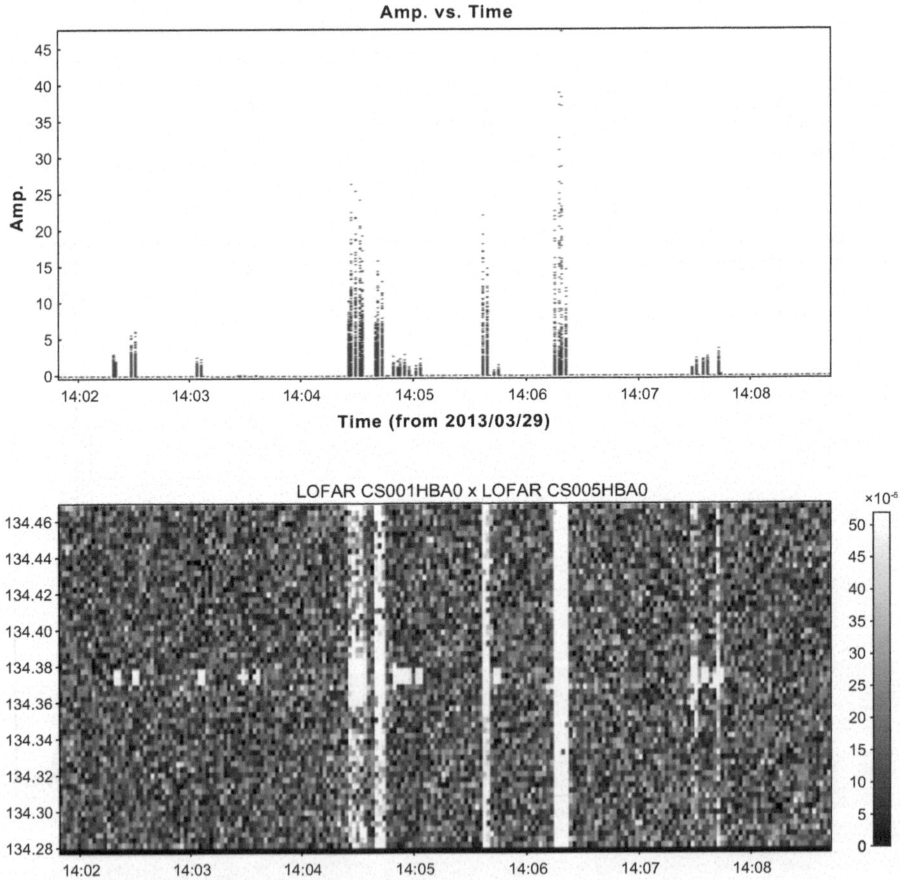

Fig. 4.2 Results from casaplotms (top) and rfigui (bottom). There are clear signs of RFI in both plots. The rfigui shows that the RFI around 14:05 is localized in frequency: only the center frequencies have to be discarded

A tool that is dedicated to visualizing RFI is called `rfigui`. This tool shows, per baseline, the amplitude of visibilities as a function of both time and frequency. This visualization is essentially also underlying the algorithms that flag data automatically.

4.3.2 *Flagging Ranges of Data (preflagger)*

When it is clear that one specific antenna was not functioning well during the observation, it is possible to flag all data from baselines involving this antenna.

Also, it is possible to flag data based on amplitude. This may seem like an effective way to remove very bright contributions from RFI, but one should keep in mind that not all RFI is very bright—e.g., also the region around RFI spikes may contain RFI at a lower level.

In DPPP, there is a step called `preflagger` that flags data based on time, baseline, elevation, azimuth, simple uv-distance, channel, frequency, or the value of amplitude, phase, real, or imaginary. The syntax for selecting specific baselines is illustrated in (Table 4.1).

Table 4.1 Baseline selection syntax

Style	Example	Description
Pattern	`CS*&RS*`	Cross-correlations between core stations (`CS...`) and remote stations (`RS...`), but not between two core stations or two remote stations
	`[CR]S*&`	All cross-correlations between core stations (`CS...`) and remote stations (`RS...`)
	`*&`	All cross-correlations
	`*&&`	All cross-correlations and auto correlations
	`*&&&`	All auto-correlations
	`0&3,4`	Baseline 0–3 and 0–4 (using antenna numbers)
Regexp	`/(RS|DE|IT).*/&&`	Cross- and autocorrelations between remote (`RS`), German (`DE`) or Italian (`IT`) stations
	`/(.*)HBA0&\1HBA1/`	Baselines between the "ears" of the same HBA station
Distance	`>1000`	Baselines longer than 1000 m
	`>5klambda`	Baselines longer than 5kλ
	`3~5klambda`	Baselines between 3kλ and 5kλ
Combination	`RS*& AND >1000`	Cross correlations between remote stations that are longer than 1000 m

This syntax works in most CASA and LOFAR tools

4.3.3 Automatic Flagging Using AOFlagger

The AOFlagger (Offringa et al. 2010, Offringa et al. 2012) analyzes, per baseline, large chunks of data and performs statistical analysis on these (refer to the journal articles for details). This analysis will work better when there is a large span of data in frequency and time. In theory, the best way to use this technique would be to perform the analysis on the whole observation at once. In practice, this will use too much memory of most computers, so there is a trade-off. If an observation is very short, the automatic flagging will perform worse as well.

4.4 Removal of Bright Sources

Another form of data corruption is caused by astronomical sources themselves. Especially in the low frequency radio sky, there are some very bright sources: CasA, CygA, VirA, TauA, HydA, HerA, together called the 'A-team', see Fig. 4.3. These sources are so bright that they effectively blind the telescope—this can be compared to sky-watching in broad daylight.[1]

Demixing is a technique to subtract the contribution of bright sources outside the field of view from your data.

To assess whether your data should be demixed, you can do the following:

- Image the data: if the data is affected by bright sources outside the field of view, large radial lines will appear whose center is the offending source.
- Follow a rule of thumb: in LBA you should always demix CasA and CygA, in HBA you should demix A-team sources that are within $\sim 30°$ of your target.
- Investigate the elevation of A-team sources during your observation. A-team sources with high elevation should be demixed. This is because, even if the signal is phase shifted by instrumental delays to any position on the sky, the individual antennas still point straight up, and are most sensitive at zenith. This can be done by using the tool `plotateamelevation.py`. See Fig. 4.4.
- Look at the amplitude vs time plot of your visibilities. Because the A-team sources will enter and leave the sensitive areas of the beam, if there are A-team sources they will lead to rising and falling of visibilities (whereas the visibilities should typically look like flat noise). This works only on longer observations.

To subtract the contribution of bright sources, the theoretical visibilities from a model of these sources is used. However, this can not be subtracted directly because the real data is also affected by unknown effects in the instrument. These unknown effects can be calibrated for (see Chap. 5) with a directional calibration. Ideally, this calibration is performed on the full-resolution data. Because this directional

[1]Indeed, the sun is also radio-bright and can be removed from day-time observations using the demixing procedure.

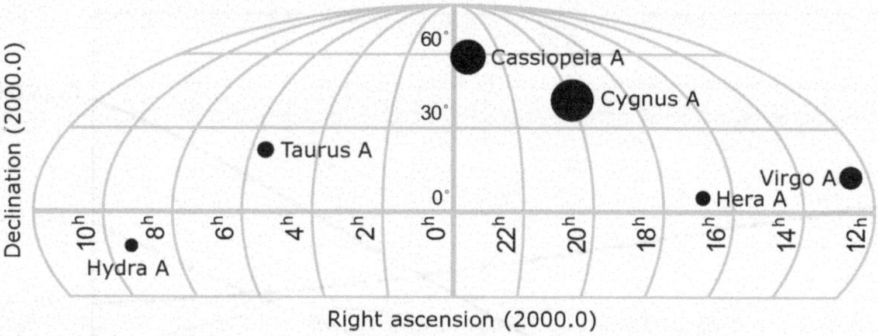

Fig. 4.3 The A-team sources that are relevant for LOFAR observations. The brightness of the sources is indicated by a larger dot

calibration on full resolution data takes a lot of computational resources, there is a technique (by S. van der Tol) that saves a great deal of computation time by calibrating on averaged data. Alternatively, faster directional calibration software exists that is currently not part of the LOFAR pipelines, see e.g. Kazemi et al. (2011) and Smirnov and Tasse (2015).

As explained in Sect. 4.1, the time and frequency resolution for data can be quite low for data near the phase center of the data. The idea of demixing is that data can be *phase shifted* to another location. If the resolution is high enough, this will yield good data quality in that direction. The demixing algorithm phase shifts the data to an A-team source, so that the fringe rate of this source is not too high. Then it averages the data temporarily in this direction, to make an efficient calibration possible and to smear the contributions from distant sources. For each A-team source, the equations for a directional calibration including all sources are set up. The equations for all directions are combined and solved, giving the calibration parameters that make it possible to subtract the sources from the data. The phase shifting, averaging and solving is handled in possibly multiple simultaneous directions, where the equations arising from the different directions are linked with a matrix called the *mixing matrix* (hence the term *demixing*).

4.4.1 Smart Demix

The task of assessing whether and which A-team sources have to be demixed has been automated in a DPPP step tentatively called `smartdemix`. The algorithm, by R.J. van Weeren, in this step simulates the A-team sources at regular time intervals. If the signal is above an empirically determined threshold, it will be demixed.[2]

Another advantage of smart demixing is that sources can be demixed for only part of the duration or a subset of the baselines of an observation, thus saving computation time for the other parts.

[2]In fact, the algorithm is more elaborate. It is documented in the DPPP documentation.

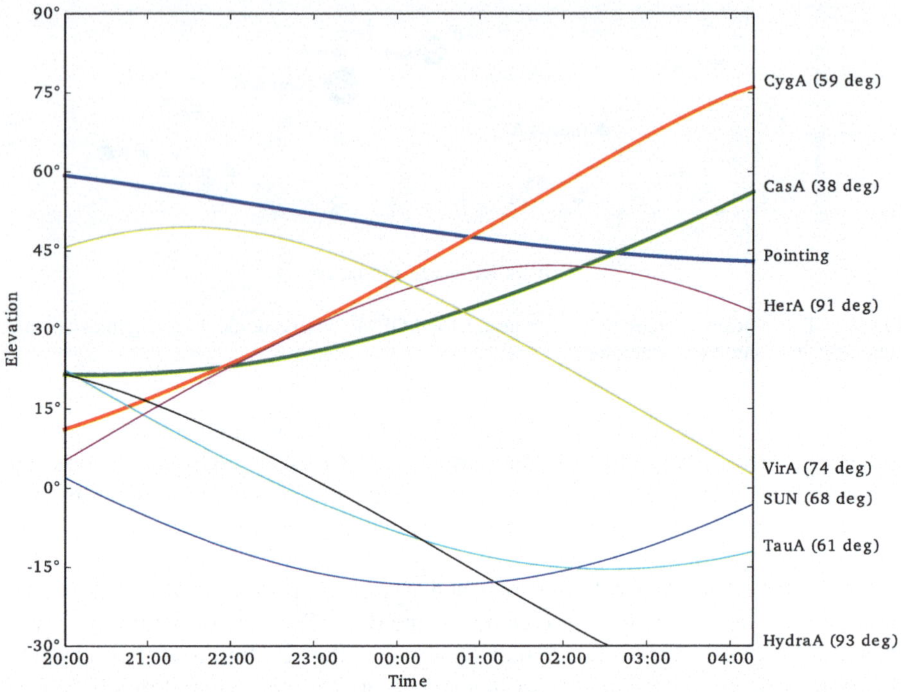

Fig. 4.4 Elevation of the A-team sources during a (short) example observation. The angles in brackets denote the angular separation between the A-team source and the pointing. In this case, Cygnus A and Cassiopeia A should be removed for the second half of the observation, because of their high elevation

4.4.2 *Averaging Data with DPPP*

DPPP enables raw data to be flagged, demixed and averaged in a sequence of steps, reading and writing the data only once. The following parset gives an idea of how this can be configured. Details about all possible settings can be found in the LOFAR Imaging Cookbook (Shulevski 2017) or the DPPP documentation.[3]

```
msin  = L114220.MS
msout = L114220_processed.MS
steps = [preflagger, aoflagger, average]
preflagger.baseline   = CS013&&  # Flag station CS013
aoflagger.memoryperc = 80        # Don't use too much memory
average.timestep     =  2        # Average two time steps
average.freqstep     = 16        # Average 16 channels
```

[3]https://www.astron.nl/lofarwiki/doku.php?id=public:user_software:documentation:ndppp.

References

Condon, J., Ransom, S.: Essential Radio Astronomy. Princeton University Press, Princeton (2007)
Kazemi, S., Yatawatta, S., Zaroubi, S., et al.: Mon. Not. R. Astron. Soc. **414**, 1656 (2011)
Offringa, A.R., de Bruyn, A.G., Biehl, M., et al.: Mon. Not. R. Astron. Soc. **405**, 155 (2010)
Offringa, A.R., van de Gronde, J.J., Roerdink, J.B.T.M.: Astron. Astrophys. **539**, A95 (2012)
Shulevski, A. (ed.): The LOFAR Imaging Cookbook (2017). https://www.astron.nl/radio-observatory/lofar/lofar-imaging-cookbook
Smirnov, O.M., Tasse, C.: Mon. Not. R. Astron. Soc. **449**, 2668–2684 (2015)
Thompson, A., Moran, J., Swenson, G.: Interferometry and Synthesis in Radio Astronomy. Wiley, New York (2008)

Chapter 5
Calibration of LOFAR

John McKean and Ger de Bruyn

Abstract Calibrating the raw visibility data from radio interferometers in order to remove the effects of the instrument and the atmosphere is needed to achieve the desired scientific outcome. Here, the processes required to calibrate data from LOFAR, after applying the pre-processing steps, are discussed. In particular, the calibration philosophy and usage of the LOFAR calibration software is discussed in detail.

5.1 Introduction

Calibration is the process that takes the raw data from a telescope and applies a set of corrections to form a science data product that can be used for a given study. In the case of interferometry at radio wavelengths, where the data are the visibilities produced from each antenna pair (baseline), the relative and absolute amplitude and phase of the raw data can be corrupted as a function of time and frequency by the instrument and the atmospheric conditions during the observation.

An example of a simple two element interferometer that measures the sky response in terms of the combined antenna voltages and the geometric delay is shown in Fig. 5.1. In the ideal situation, the signals arrive at the two antennas after some relative delay, which is equivalent to the geometric delay due to the relative positions of the antennas with respect to the distant source. This delay can be compensated in a particular direction (called the delay tracking centre). The voltages produced at each antenna are then multiplied and time-averaged to

J. McKean (✉)
ASTRON, Dwingeloo, The Netherlands

Kapteyn Astronomical Institute, University of Groningen, Groningen, The Netherlands
e-mail: mckean@astron.nl

G. de Bruyn
ASTRON, Dwingeloo, The Netherlands

G. Heald et al. (eds.), *Low Frequency Radio Astronomy and the LOFAR Observatory*, Astrophysics and Space Science Library 426,
https://doi.org/10.1007/978-3-319-23434-2_5

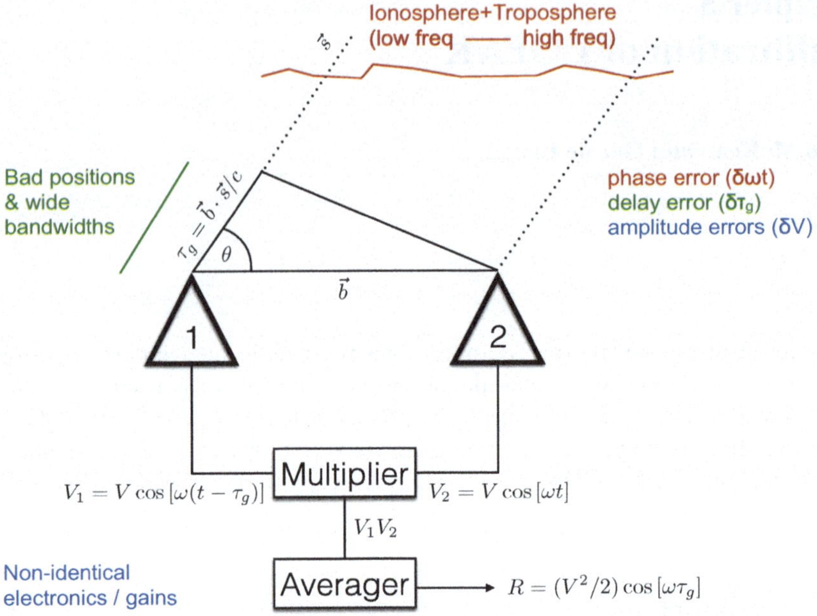

Fig. 5.1 Schematic example of the response (real part) of a two-element interferometer, consisting of two antennas separated by **b**, pointing towards a source in the direction of unit vector **s**. In the ideal case, the radiation from the source arrives at antenna 1 after some time-delay, defined by the geometric delay τ_g, which results in a phase shift relative to antenna 2. The voltages from the two antennas are multiplied and averaged within the correlator to produce the response function R to the sky emission. However, the atmosphere can add time dependent phase shifts, poor antenna positions and wide-bandwidths add frequency dependent phase shifts and the electronics (e.g., amplifiers/mixers) in the receiver system can add time dependent amplitude and phase shifts. Calibration is used to correct for these corrupting elements within the observed response R

produce the response of the interferometer towards some direction. However, reality often differs from this ideal situation. The atmosphere is not a stable uniform slab, but has small scale-variations that change the effective path length that the signals traverse towards each antenna. This results in a slight change in the delay compared to what we expect, and therefore, an apparent change in the position of the object in the sky. Also, the extra path length that is corrected for by the geometric delay is particular for a given frequency and direction in the sky, which is corrected for using a delay model that takes into account the relative antenna positions during correlation. However, any small error in this delay model, coupled with observations that are made with a wide bandwidth will show a different delay for the bottom and top end of the frequency band. The time and frequency dependence of the delay, due to the atmosphere and the bandwidth of the observation, can corrupt the phase information of the signals and result in a reduction in the measured correlated signal. Furthermore, the ideal case expects that the amplitude of the incoming signals are the same. However, these can also be affected by the atmosphere (to some extent), but mainly due to the differing amplifiers, electronics or directional response of the individual antennas. In the case of the atmospheric and system variations, the change in the amplitudes tend to be quite small (although this is dependent on the observing wavelength). The largest variations come from the direction dependent response of the antennas.

The standard method that is used to correct for these variations in the amplitude and phase as a function of time and frequency is to solve the Radio Interferometric Measurement Equation (RIME; Hamaker et al. 1996; Smirnov 2011). This equation relates the observed visibilities produced by combining antennas i and j ($\mathbf{V}_{ij}^{\text{obs}}$) to the expected visibilities ($\mathbf{V}_{ij}^{\text{model}}$; determined from some model for the sky) via a set of 4×4 Mueller matrices ($\mathscr{J}_{ij}$),

$$\mathbf{V}_{ij}^{\text{obs}} = \mathscr{J}_{ij}\mathbf{V}_{ij}^{\text{model}} \tag{5.1}$$

and where the notation for the Mueller matrices relates to the calibration solutions for the individual antennas,

$$\mathscr{J}_{ij} = J_i \otimes J_j^{\dagger}. \tag{5.2}$$

These solutions, often referred to in practice as calibration tables, are represented by a set of 2×2 Jones matrices (Jones 1941), where the individual elements of the matrices describe the antenna based calibrations, for each correlation, for a given correction (gain, bandpass, delay, etc.). For example, the gain solutions for the two linear polarisations X and Y are defined as,

$$J_i^{\text{gain}} = \begin{pmatrix} g_X & 0 \\ 0 & g_Y \end{pmatrix}, \tag{5.3}$$

and the leakage terms are defined as,

$$J_i^{\text{leakage}} = \begin{pmatrix} 1 & D_X \\ D_Y & 0 \end{pmatrix}. \tag{5.4}$$

The advantage of using this Jones matrix formalism is that various calibration tables can be easy combined linearly to determine the overall corrections to the data,

$$J_i^{\text{overall}} = J_i^1 J_i^2 J_i^3. \tag{5.5}$$

As Eq. (5.1) shows, the calibration process is dependent on comparing the model for the sky with the observed visibilities to determine the calibration solutions. Therefore, any error in the sky model can be compensated, to some extent, within the calibration tables. However, as will be discussed below, the process of self-calibration can be used to iteratively improve the model for the sky, and hence to determine a more accurate set of calibration tables. As Eq. (5.1) is invertible, we can use these tables to correct the observed visibilities. The calibration solutions are determined via a process that is called *closure*; that is, the expectation the phase (ϕ_{ij}) and amplitude (V_{ij}) of groups of baselines have certain properties given the source structure. The closure phase between three independent antennas is,

$$C_{ijk}(t) = \phi_{ij}(t) + \phi_{jk}(t) + \phi_{ki}(t), \tag{5.6}$$

where $\phi_{ki}(t) = -\phi_{ik}(t)$ because visibilities are Hermitian. For a point source at the phase centre the closure phase has the property that $C_{ijk}(t) = 0$. The amplitude closure between four independent antennas is given by,

$$\Gamma_{ijkl}(t) = \frac{|V_{ij}(t)||V_{kl}(t)|}{|V_{ik}(t)||V_{jl}(t)|}. \tag{5.7}$$

Note that closure is invariant to additive antenna-based phase errors, and multiplicative antenna-based amplitude errors. As suggested by Eqs. (5.6) and (5.7), for a large number of independent antennas, these solutions are well constrained since the number of unknowns is N_{antennas} and the constraints are $N_{\text{baselines}} = N_{\text{antennas}}(N_{\text{antennas}} - 1)/2$.

An example of applying the Measurement Equation to observed data is shown in Fig. 5.2. In this example, the observed phase, amplitude and delays are compared with the expectations for a point source model with a flat radio spectrum (which in this case is appropriate). For this model, the phase, amplitude and delays are solved to determine the corrected (calibrated) data. Taking the phase first, we see that there are two issues with the raw data; the relative phases of each antenna pair differ significantly from 0° due to the electronics of the receivers, and there are small-scale variations over time due to the atmosphere. The calibration process corrects for both the relative offset and the small-scale variations as a function of time so that the final phases are within ±5° of the expected value. The scatter is due to the noise of the visibilities. The delays show a slope in the phase as a function of frequency due to the large fractional bandwidth of the data set in this example. The calibration process fits for this slope and removes it from the data, giving a flat spectral response of the phase. Note that after correcting for the delay, the phase information can be averaged in frequency to increase the signal-to-noise for the calculation of the temporal phase variations. Finally, the amplitudes in this example show a large relative variation, with an offset from the expected absolute flux density of the object. These variations are dominated by the differing receiver sensitivities (longer time-scale data may also show a variation due to the differing directional response of the antennas). The calibration process corrects for the relative and absolute variations, with the scatter due to the noise of the individual visibilities.

The example shown in Fig. 5.2 is for a simple point source that dominates the sky brightness distribution. But, for arrays with a large field-of-view, like LOFAR, the model visibility function is always more complex. Below, we will discuss how to construct a good estimate of the model sky, but the basic process of solving the Measurement Equation is the same. This chapter deals with the specific issues

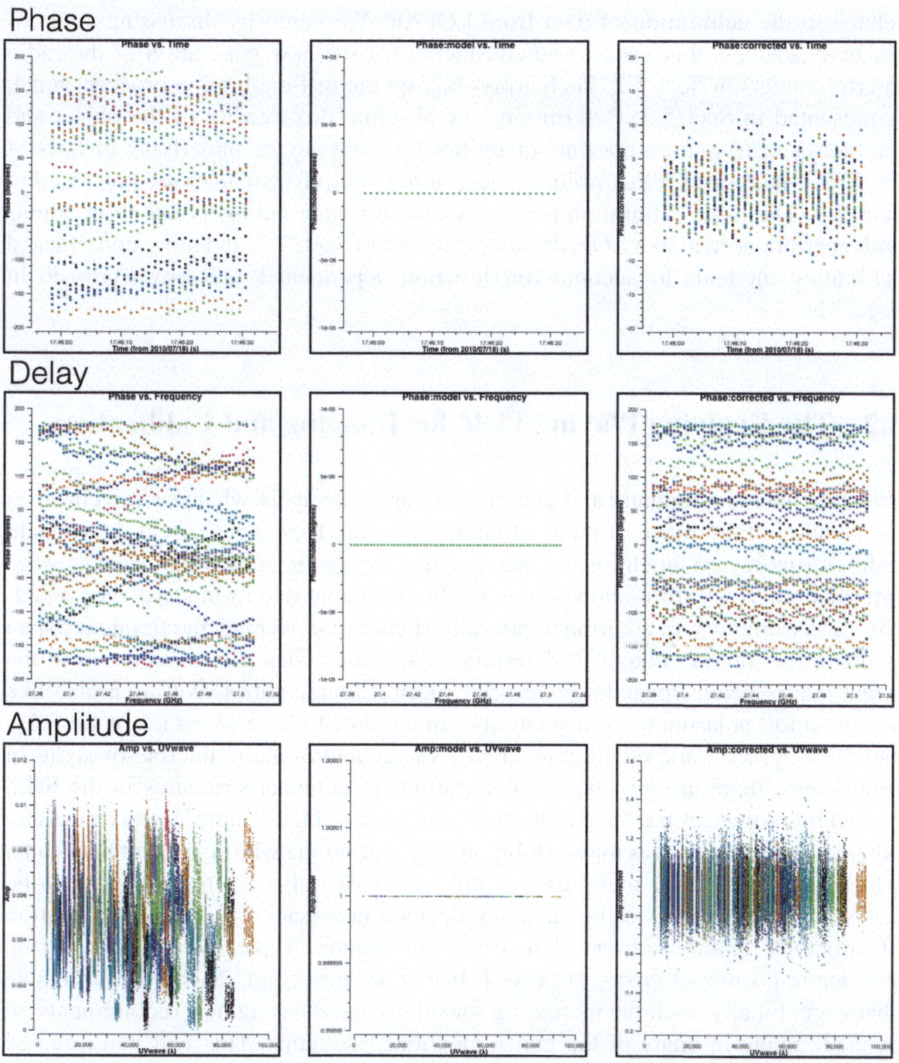

Fig. 5.2 Example of the calibration process for the phase (upper), delay (middle) and amplitude (lower) using a point source model (an object with a constant phase of 0° as a function of time and frequency, and a constant amplitude on all baselines) located at the phase tracking centre. The data are for visibilities produced from 1 antenna to 26 other antennas, with the colour showing the different antenna pairs. For each example, the raw data (left), the model data (middle) and the calibrated data (right) are shown. Note that the calibrated data differ from the model due to the noise of the individual visibilities, and pay attention to the vertical scale in the phase plots

related to the calibration of data from LOFAR. We begin by discussing some of the new concepts that need to be considered for the new generation of the radio interferometers in Sect. 5.2. Such arrays rely on the self-calibration process, which is presented in Sect. 5.3. Determining the absolute flux-density of the object and the radio spectrum are important quantities for studying the underlying physics of the target; methods for determining these at low radio frequencies are presented in Sect. 5.4. The basic calibration processes used for wide field-of-view observations with aperture arrays, like LOFAR, are presented in Sect. 5.5, and more complicated calibration methods to account for direction dependent issues are discussed in Sect. 5.6.

5.2 The Evolving Playing Field for Imaging and Calibration

Probably the most fundamental goal of an imaging array, in whatever waveband, is the creation of an image of the continuum emission. However, current demands in radio astronomy go much further than just making an image; we also want spectral information. This information is now readily available due to modern very broad-band spectrometers. In addition to providing better uv-coverage, the frequency span is often also broad enough to determine a spectral index distribution across the target that is being imaged; so we need spectral image cubes. We often also like to extract full polarimetric information from our data to learn about magnetic fields and the magneto-ionic medium inside the source and/or along the line-of-sight. In some cases, there are also temporal variations in compact structures in the field, caused by propagation effects induced by the interstellar or interplanetary medium. Alternatively, the sources may exhibit intrinsic variations within the synthesis time, which can extend over many days if not weeks. We also desire our image to be astrometrically correct to the accuracy deemed necessary for cross identification of structural features observed in other wavebands. Typically we demand the systematic positional errors to be well below an arcsecond. This is a formidable challenge. Finally, with the increasing sensitivity have sprung new requirements on dynamic range in many of the above mentioned domains. This will be discussed later in this chapter.

Thus, our data and the required information live in a five-dimensional space: two spatial coordinates, one frequency dimension, one temporal dimension and one for polarization (linear and/or circular). Note that within the solar system, and certainly in the terrestrial environment, we might even contemplate determining distances of the emitting sources as well! The treatment of polarisation information is discussed in more detail in Chap. 10 and so we will not discuss it here, except insofar as it touches upon the calibration of our data. Similarly, time-domain science (e.g. pulsars, transients, etc.) is a very rich field of research requiring special observing modes and calibration techniques, and these are discussed in Chap. 14. However, slow temporal variations of compact sources in the target field, whether due to intrinsic or extrinsic causes, will occasionally complicate the calibration of the data. How to find such sources, and recognise their effects, is not always easy.

An example of a source that varied dramatically within a 12 h synthesis time is the quasar J1819+3845 (Macquart and de Bruyn 2007). This example is, however, unusual because the brightest source in the image was also the most dramatically variable. Removing its time-variable flux density was required to reach the thermal noise level and reveal the larger population of fainter sources in the field. At the low frequencies where LOFAR works, the brightest sources are unlikely to be highly variable, but there will be variable sources that might cause some low-level residuals if one is not aware of their variability.

5.3 From Calibration to Self-Calibration

5.3.1 The Self-Calibration Philosophy

As has been described above, a radio telescope needs to be calibrated for a variety of instrumental parameters. For terrestrial radio astronomy we also need to worry about the corrupting effects of the atmosphere. Standard calibration uses a bright calibrator source (one with a well-defined source structure at the frequency and resolution of the observations) to determine the calibration solutions of the Measurement Equation as a function of time and frequency (see Fig. 5.2). These calibration solutions are then applied (transferred) to the target field that is observed in close spatial and temporal proximity to the calibrator source. However, this process can only create images with a limited dynamic range. Depending on the brightness of the sources within the field-of-view this may already be sufficient. For example, in the case of LOFAR HBA observations, an image made using only the core stations will already have a dynamic range[1] of a few × 100:1 or more, which is sufficient for a quick assessment of the science-quality of the data set. However, this is usually not good enough for the majority of the science goals that the data set was intended to tackle. Fortunately, the target source and other sources in the field, if sufficiently bright, can be used to recognise, model and remove remaining errors. Knowledge and experience with how certain types of errors in the data manifest themselves, through the Fourier transform of these corrupted data, into the image will come in very handy at this stage. This requires experience and platting with various types of errors (see Chap. 6). Using the sources within the field itself to calibrate the remaining errors, and improve the dynamic range in the image, is called self-calibration and this technique has allowed the quality of radio astronomical images to be improved significantly in the past 35 years.

[1]The dynamic range is the ratio between the peak brightness and an estimate of the noise in an image. High-quality images have high dynamic range.

The standard method of self-calibration uses an iterative cycle of imaging and calibration to determine an improved model for the sky and hence better calibration solutions and calibrated data. The cycle has five basic steps.

1. Determine an initial model for the observed sky surface brightness distribution (via making an image of the initially calibrated data or from published sky catalogues).
2. Use this model to calibrate the relative gains (amplitude and phase) over some solution interval by applying the Measurement Equation.
3. Determine an improved model for the observed sky surface brightness distribution by making an image of the calibrated data.
4. Use this model of the observed sky surface brightness distribution to self-calibrate the relative gains (amplitude and phase) over some solution interval by applying the Measurement Equation.
5. Iterate over steps 3 and 4 until no further improvement in calibration and imaging is made.

This method of self-calibration has the advantage of correcting for residual amplitude and phase errors produced by applying an inadequate starting model (e.g. from a previous sky survey) that can be taken at a different time, frequency and resolution to the target data set. Furthermore, if the solution transfer from an out-of-beam calibrator source is used to determine the initial relative gains, then there are likely to be small residual errors due to direction dependent effects. A better model for the observed sky surface brightness distribution can also be used to solve for direction dependent effects within the wide field-of-view of a typical LOFAR data set (see below).

The disadvantage of the self-calibration method is that the calibration solutions will only be as good as the model that is used, as can be seen from Eq. (5.1). Therefore, errors in the initial model can propagate into the self-calibration solutions and it is possible for the process to diverge from the correct model for the sky. A classic example of this is when very low resolution data, for example with beam size around 2 arcmin full width at half maximum (FWHM), is used to calibrate the core and remote stations of LOFAR. For the core (baseline lengths $\leq$ 3 km), the effective resolution of the data is similar to that of the model and good phase and amplitude solutions are typically obtained. However, for remote stations (baselines 3–85 km) where the effective resolution is $\geq$ 3 arcsec, the situation can be significantly different. Here, the sources are likely resolved into several components, which has two effects. First, the source positions will change due to the structure of the sources, which will affect the phase solutions and second, the total correlated flux will typically decrease on the longer baselines, which will effect the amplitude solutions. An example of this is shown in Fig. 5.3, where a simple model has been used to calibrate the core and remote baselines for the complicated radio source Cygnus A. In this case, after several iterations of self-calibration to generate a more accurate model for the sky, better solutions are obtained. Although the large number of stations increases the unknowns (we solve for the amplitude and phase per station as a function of time and frequency), we also dramatically increase the number of

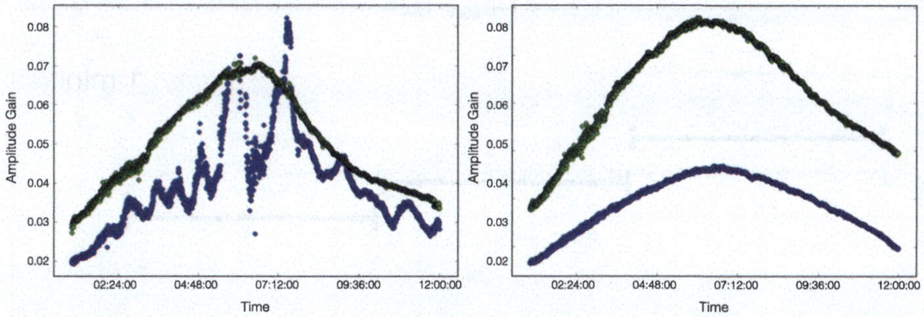

Fig. 5.3 Left: Example of the amplitude solutions obtained for a core (green) and remote (blue) station in the case of an observation of Cygnus A and using a simple model (and without a correction for the varying primary beam). The core station shows the expected, smoothly varying amplitude solutions given the beam response of the telescope, whereas the solutions for the remote station show significant small time-scale variability, which is due to there being resolved structure on remote station baselines that is not accounted for in the model. Right: The solutions after several iterations of self-calibration. Now both the core and remote stations show the expected smooth variation over time. Note that the two types of stations have different relative amplitude gains due to their different collecting area

constraints, as the number of baselines increase to $N_{\text{antennas}}(N_{\text{antennas}} - 1)/2$. For example, the 48 sub-stations of the LOFAR core have 96 unknowns (amplitude and phase), but 648 constraints from the baselines, making the self-calibration process a well constrained problem.

It is also important to have sufficient baseline sensitivity when determining the calibration solutions, otherwise the solutions themselves will become noisy. As a rule, it is often the case that the baseline sensitivity should have a signal-to-noise ratio of > 3. This can be achieved if there is sufficient correlated flux on all of the baselines, or alternatively by choosing a solution interval that is long enough to obtain the required sensitivity (recall that the sensitivity improves with $1/\sqrt{\text{time}}$). On the other hand, if the solution interval is too long, then it will not be possible to correct for small time-scale phase-variations; see for example Fig. 5.4. Ideally, we want to choose a solution interval that is short enough to track the phase and/or amplitude variations, while maintaining a sufficient signal-to-noise ratio on each baseline.

Therefore, the choice of solution interval is very much dependent on the ionospheric observing conditions and the signal-to-noise ratio of the data, in the case of the phase variations. In the case of LOFAR, solution intervals for phase self-calibration of a few to 10 s is typically sufficient. In the case of the amplitude self-calibration, the amplitude variations due to the LOFAR beam response will be varying much less with time, and solution intervals of around 10 min tend to be appropriate.

Overall, the self-calibration process has become the method of choice for determining the most accurate calibration solutions for a given data set in radio astronomy. An example of the dramatic difference in image quality before and after self-calibration is shown in Fig. 5.5.

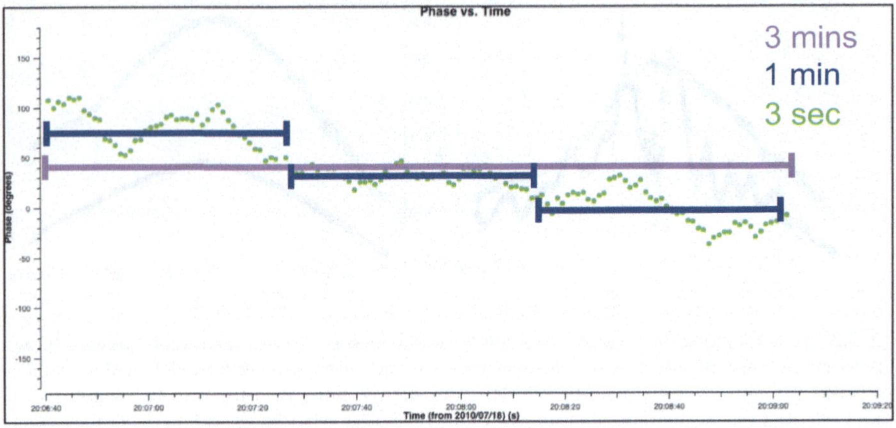

Fig. 5.4 Example of the phase variations seen over 3 min for a point source, that is, a source that should have a constant phase of 0° over the 3 min time interval. For a solution interval of 3 min, the phase-offset of 40° can be corrected for, but there will be maximum residuals of ±60° over the time period. Taking 3 × 1 min solution intervals (giving three solutions) improves the maximum residuals to ±30°. Finally, if the signal-to-noise ratio is sufficient, we can use the visibility integration time as the solution interval (60 solutions)

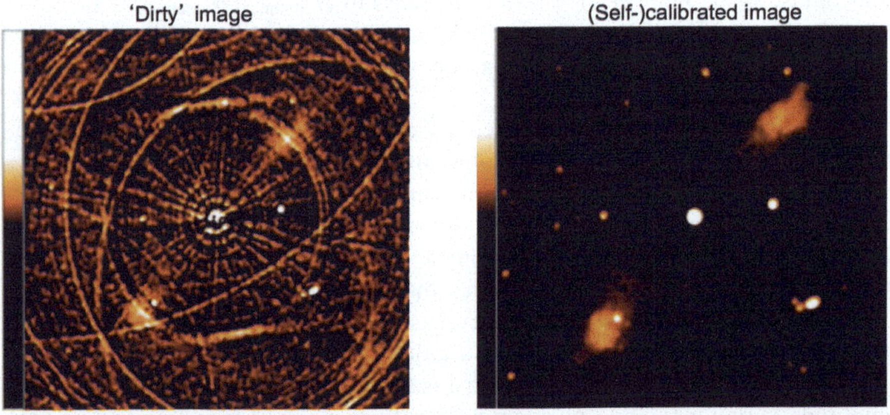

Fig. 5.5 Images of the giant radio source B1245+67 as observed with the WSRT in 1991 (de Bruyn, unpublished). Shown are a dirty image, following standard calibration, and partial deconvolution (left), and a self-calibrated image (right). The central source of this double-double radio galaxy, which dominates the error patterns in the image, has a flux density of 250 mJy, about 5000 × the rms noise level

5.3.2 Calibration Regimes and Unknowns

Conceptually the calibration of an aperture synthesis array like LOFAR can best be divided into three distinct parameter regimes dealing with (1) the sky, (2) the ionosphere and (3) the instrument. These three groups of unknowns each have their distinct timescales, allowing them to be separated relatively easily. The sky is usually the most stable of the three, as discussed above. The instrument, if well designed, can also be assumed to be stable, i.e. deterministic, down to some level of accuracy. The ionosphere on the other hand, can corrupt data on timescales as rapid as seconds. This is the case during observing conditions where the ionosphere will show small scale structure, smaller than the Fresnel scale at the height of the ionosphere (typically 1 km; see Vedantham and Koopmans 2015). Fortunately, the more typical timescales are tens of seconds to minutes. How we go about determining the parameter values for these three classes of unknowns will be discussed in the following subsections.

5.3.2.1 Modelling of the Sky

We need an approximate model of the sky within the field-of-view, to start self-calibration. When we start our calibration and imaging process we usually do not know in detail what the sky looks like. Images or source catalogues from previous surveys, preferably at frequencies not too different from the data to be calibrated, can be used to kickstart the calibration. This is sometimes referred to as the starting or initial sky model. Data from Westerbork Northern Sky Survey (WENSS; Rengelink et al. 1997), taken at 325–350 MHz, and the Very Large Array Low-frequency Sky Survey (VLSSr; Lane et al. 2014), taken at 74 MHz, can and have been used for the initial sky model of LOFAR data (depending on the frequency and angular resolution of the data set). The recent TIFR GMRT Sky Survey Alternative Data Release (TGSS-ADR1; Intema et al. 2017) is now an excellent resource to generate a starting calibration model, and at declinations below about $\delta \sim +30°$ the MWA's GaLactic and Extragalactic All-sky Murchison Widefield Array (GLEAM) survey (Hurley-Walker et al. 2017) provides a very useful sky model at relatively low angular resolution, but with excellent frequency coverage. Although the calibration of the extremely large volume of data (typically tens of TeraBytes per long-track observation) will take a large amount of compute resources, which can extend over many weeks if not months, it is important to realise that for the astronomer most of the time will go into developing a strategy to construct a good sky model. It is therefore important to develop a strategy that will permit one to realise the science goals. For example, if accurate astrometry is required (see next subsection) ionospheric effects should be taken into account from the start in the construction of an astrometrically precise sky model. In any case, the sky model that will be generated should be carefully saved. It will allow the astronomer to recreate good images from the raw data within a small fraction of the time that was needed to generate this sky model.

Fortunately, the LOFAR core, with 24 (LBA) or 48 (HBA) stations, provides excellent uv-coverage to generate low-resolution images, even within a snapshot interval of 10 s. The LOFAR core is operated in an instrumentally phase-stable way (using a single time and frequency standard), and the ionosphere is usually fairly stable on the short baselines (<1000 wavelengths) as well. The initial (raw) image made with only the core stations therefore will provide an image with a dynamic range of about 100:1. This is sufficient to start the self-calibration process. One usually needs an external calibrator, preferably within the same station beam, to set the absolute flux (see below). Using data from the LOFAR Multi-frequency Snapshot Sky Survey (MSSS; Heald et al. 2015) or TGSS-ADR1 a network of position and flux calibrators is now available for baselines out to ~3 km (i.e. the LOFAR core) and beyond. When the full Dutch LOFAR is used, with baselines out to ~85 km, ionospheric seeing is a major complication. All sources will then be smeared ('convolved') with a time-variable seeing disk that will complicate modelling the fine-structure in the source, which is particularly important to calibrate the remote stations of LOFAR. This almost looks like a catch-22 situation. An iterative approach has therefore been developed where imaging and self-calibration cycles are employed in steps, where the baseline length of the uv-data is increased iteratively (see Chap. 9 for a discussion of the LOFAR imaging and self-calibration pipeline). The same reasoning holds if the international stations, with baselines exceeding 1000 km, are part of the array, and therefore must be part of the model-building exercise.

5.3.2.2 Measurement of Positions

The creation of images with accurate positions can vary from straightforward to being a very time-consuming task. This all depends on the desired accuracy. Here we must also make a distinction between absolute and relative positions. Getting absolute positions is very difficult. However, positions relative to a global network of sources (like MSSS) is usually sufficient for most science goals. That is to say, someone else has already done the hard work for you in the past! However, for cross identification with images in other wavebands we require positions to arc second accuracy. This is far from easy at low frequencies. This is due to the corrupting influence of the ionosphere, which can move sources around over tens of arc seconds at a frequency of 100 MHz. At the low frequencies where LOFAR operates, ionospheric refraction can have a very large effect.

The refraction angle scales with wavelength squared and also scales linearly with the path integral of the number of electrons in the ionosphere. This integral is known as the total electron content (TEC; see Chap. 7 for details). At (geographic) mid-latitudes, where LOFAR operates, the ionosphere can have a vertical TEC (VTEC) of 20–30 TEC units during daytime. At an elevation of 30° this would approximately double. On the other hand, typical VTEC values for nighttime observations are only about 5 TEC units. The differential refraction for various frequencies and elevations, for a plane-parallel ionosphere in a curved Earth setting, is shown in Fig. 5.6 for a VTEC of 5 TEC units.

Due to the large size of the station beam, as compared to arrays working at higher frequencies, LOFAR is sensitive to emission from sources separated by up to 10° apart. This implies that there will be a significant amount of differential refraction, especially for sources that differ in elevation (which for observations

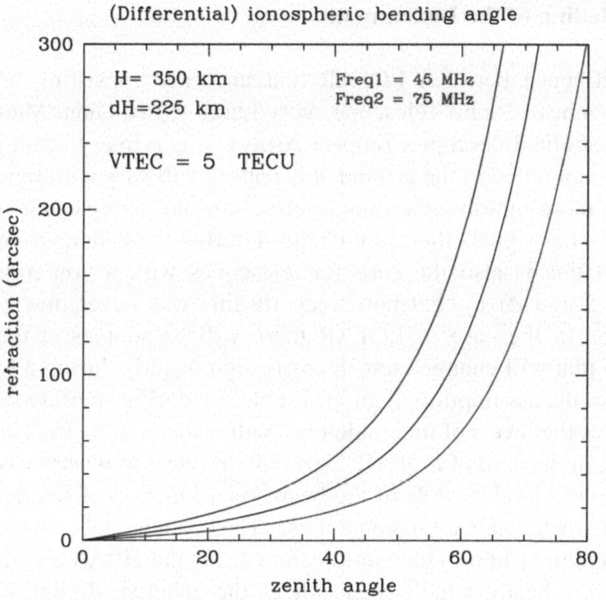

Fig. 5.6 The ionospheric refraction angle at two frequencies with the LOFAR HBA band as a function of zenith angle. The differential refraction between these two frequencies is shown in black. The VTEC was assumed to be 5 TEC units (TECU), a typical night-time value for the Netherlands. The refraction is not very sensitive to the precise height and thickness of the ionosphere, which are also given

at transit corresponds to declination). Depending on the observing frequency, the VTEC could easily amount to many arcseconds if not arc minutes of differential refraction. To exacerbate the effects of differential refraction is that these effects will also change with time during a long synthesis. To properly deal with differential refraction we need to have a model of the ionospheric 'phase-screen'. For arrays of large physical size (100–1000 km) and for projects that involve wide field imaging, we may well need a 3-D model. How far we have come in modelling ionospheric phase screens is described in Chap. 7. If we could construct such a phase screen, then we also need a method to apply these corrections as a function of time. This can in principle be done using the AWImager (Tasse et al. 2013, see Chap. 8) or WSCLEAN (Offringa et al. 2014), where both time-variable station beams and ionospheric delay effects can be corrected for. However, although the technique has been demonstrated with MSSS data (Heald et al. 2015), it is in general not computationally feasible at the time of writing.

5.3.2.3 Modelling of the Instrument

The main difference between LOFAR and an array consisting of dishes (e.g. Westerbork Synthesis Radio Telescope, Very Large Array, Giant Metrewave Radio Telescope, Australia Telescope Compact Array) is its primary beam pattern. For a phased-array station tied to the ground, this pattern will vary with time; slowly near the zenith but more rapidly when lower elevations are approached. The reception pattern for celestial signals therefore changes during the course of an observation. At some level this is also the case for telescopes with a non-equatorial mount like the Very Large Array, but until recently this was never much of a concern for most users. In the case of LOFAR there will be sources at the edge of the primary beam that will change their intensity significantly during a long synthesis. This invalidates the assumption of an invariable sky during aperture synthesis. With variable sources the levels of their sidelobes will change. The effect is indeed much more dramatic in the case of LOFAR, especially if one also wishes to observe down to low elevations. Good models of the beam, as a function of frequency and time, have been constructed and are in general use. The beam model contains two to three components: (1) the (dipole) element beam; (2) for the HBA only, the tile beam[2]; and (3) the array beam, which encapsulates the coherent digital addition of the signals from either LBA dipoles or HBA tiles in one or more directions. These beam models, available in most LOFAR software tasks, might suffice if the components do not change with time. Unfortunately that is not the case. A small, but non-negligible fraction of the HBA tiles in the core stations of LOFAR do not work optimally and need occasional repair. Keeping track of the active elements is therefore crucial. This is important in two places: (1) when digitally adding signals and (2) when computing the station array beam for calibration purposes. The elements that are used for any given observation will therefore change the beam pattern on the sky. The information about which elements are operational during an observation is now recorded within the Measurement Set metadata and so can be used to modify the calculated beam pattern.

An important aspect of a phased array with a regular arrangement of elements, as is the case for the HBA antenna stations, is the occurrence of grating lobes. The critical frequency of a LOFAR tile is 120 MHz, by design. As soon as one observes at higher frequency, and/or at low elevation, the telescope becomes sensitive to signals from more than one direction. This not only results in a loss of gain (sensitivity) in the target direction, but signals will also be picked up from other directions in the sky, which will not be optimal. All LOFAR observers are familiar with the latter. The sources Cassiopeia A and Cygnus A (intrinsically close to 10 kJy at 150 MHz) are almost always corrupting the faint signals from the target field. However, when grating lobes with an intensity approaching the main lobe start to

[2]For the HBA, each tile beam is formed from 16 dipoles within each tile. Here, an analogue beamformer is used at the tile level to combine these dipoles to form a tile beam that is controlled at station level.

appear, many sources sufficiently bright to be catalogued in the 3C (Laing et al. 1983) or 4C (Pilkington and Scott 1965; Gower et al. 1967) catalogues will start to add their signals to the visibilities. They can often be recognised through the rapidity of the fringes. Therefore, high frequency (>160 MHz) observations at low elevation should be avoided, if at all possible. However, if this can not be avoided, the sky modelling task becomes much more elaborate. Several filtering techniques (van der Tol et al. 2007; Offringa et al. 2012) have been developed to reduce the effect of sources far from the phase centre. These typically entail solving for the contribution of the off-axis sources to the visibility function and then subtracting them.

5.4 Flux-Density and Spectral Bandpass Calibration

Having determined the relative amplitude gains of the different stations it is clearly useful to convert these to an absolute flux density scale as a function of frequency. This allows the astronomer to derive physical quantities (e.g. luminosity, brightness temperature) that are intrinsic to the object or sample of objects of interest. This process also removes the final imprint of the instrument to the data since the relative gains of the receiving system can vary from day-to-day.

The standard method for determining the absolute flux-density scale as a function of frequency is to observe an object of known flux density and to compare the relative gains between this object and the target field of interest. This so-called bootstrapping calibration method has been extended to LOFAR frequencies by Scaife and Heald (2012), who determined the spectral energy distributions between 30 and 300 MHz for six bright sources from the 3C survey over the full 0–24 h in right ascension, and tied them to a common flux-density scale (Roger et al. 1973) that avoids problems at the lowest frequencies related to secular variability of Cassiopeia A in the more common (Baars et al. 1977) flux scale. The spectral energy distributions of these sources are shown in Fig. 5.7 and are fitted with a polynomial function of the form,

$$S[\mathrm{Jy}] = A_0 \prod_{i=1-3}^{N} 10^{A_i \log_{10}^{i}(\nu/[150\,\mathrm{MHz}])}, \tag{5.8}$$

where $S[\mathrm{Jy}]$ is the flux density in Janskys, ν is the frequency (normalised to 150 MHz) and A_i are the polynomial co-efficients. This flux scale has since been linked to that at higher frequencies to yield a scale valid over three orders of magnitude in frequency (Perley and Butler 2017).

Although the radio spectra of this set of calibrators are well defined, the accuracy at which the calibration transfer is applied is clearly useful to know. It is likely that the flux-density calibrator used in an observation is in a different part of the sky than the target field. As has been discussed above and is demonstrated in Fig. 5.3, the beam of each station is not constant with position, but varies strongly with

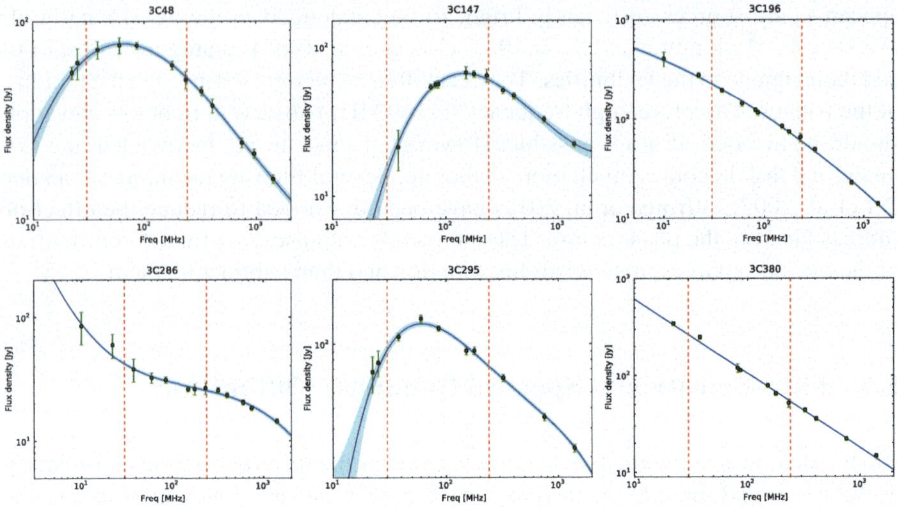

Fig. 5.7 The radio spectral energy distributions of the six compact primary flux-density calibrators used for LOFAR. The data points, shown in green, are taken from literature measurements and the best fit models (and the uncertainties) are shown by the blue regions. The red vertical dashed lines mark the regions in frequency space where LOFAR operates. Adapted from Scaife and Heald "A broad-band flux scale for low-frequency radio telescopes" MNRAS (2012) 423, 30 by permission of the Royal Astronomical Society

source elevation. Also, due to the large-bandwidths of a typical LOFAR observation, the beam will also change with frequency (recall that the beam full-width at half maximum of a station with diameter D is $\sim \lambda/D$). This means that unless the beam of each station is known sufficiently well, the relative gains between the calibrator and target field may not be well determined, which will introduce an absolute flux-density calibration error that will also vary as a function of observing frequency. In practice, the transfer of amplitudes from calibrators to target fields over arbitrary distances on the sky still results in a fairly uncertain flux scale. Therefore, an in-field flux scale bootstrapping procedure is commonly used to align with flux densities from a collection of reference surveys; see Hardcastle et al. (2016).

5.5 Calibration Procedure

As has been discussed above, LOFAR is a radio interferometer where the incoming signal is corrupted by the intervening atmosphere and the properties of the instrument. This results in a change in the amplitude and phase of the visibilities as a function of time and frequency.

In terms of the phase (ϕ), the change over time (t) and frequency (ν) are referred to as the *rate* and the *delay*, given by,

$$\frac{d\phi}{dt} = \text{Rate}, \tag{5.9}$$

and,

$$\frac{d\phi}{d\nu} = \text{Delay}. \tag{5.10}$$

Phase errors in the data will result in a loss of coherence and the image quality can be significantly reduced; recall that the geometric delay corrects for the phase difference of the incoming signal in some direction, and so, phase errors can be considered as the interferometer receiving signals from a different apparent direction. Similar to the effects of seeing in optical astronomy, these phase errors result in the object moving position on the sky, which will either completely blur the source out in the worst case, or result in the source being slightly distorted in the best case. The latter will impact the final dynamic range of the image, which is discussed below. For this reason, phase errors are typically not symmetric when inspected in the image plane (see also Chap. 6).

Amplitude errors are also a function of frequency and time due to the large bandwidths and the position dependent beams used in LOFAR. As the response of an interferometer as a function of position on the sky is a sinusoidal function, an amplitude error manifests itself as the side-lobe structure having a larger (or smaller) intensity, when compared to the case of perfectly calibrated data. This results in the side-lobe structure of sources in the sky being incorrectly modelled during imaging, and therefore, amplitude errors have a symmetric structure when seen in the image plane, as also discussed in Chap. 6.

We correct for these amplitude and phase errors by solving the Measurement Equation, given a model for the sky surface brightness distribution. In this section, we introduce and discuss the usage of the Black Board Self-Calibration (BBS; Pandey et al. 2009) software system, which was specifically developed for calibrating LOFAR data. Here, only a summary of the calibration processes and philosophy within BBS is given. For a detailed discussion of the usage and suggested input parameters, the reader should consult the LOFAR Imaging Cookbook.

5.5.1 *Black Board Self-Calibration: An Introduction*

BBS is a software package that has been designed for the calibration and simulation of LOFAR interferometric data. Although there are several traditional reduction packages available for the analysis of data from radio interferometers, a new package was needed because:

1. the datasets from LOFAR are typically very large (for example, a 6 h observation can easily generate a 100 TB dataset), and so the processing needs to be carried out in a distributed fashion, using dedicated compute clusters;
2. the beams of the LOFAR stations are variable with position on the sky, as well as the number of operational elements within each station that were used in the observation; and,
3. due to LOFAR's large field-of-view, calibration solutions need to be determined and corrected for in multiple simultaneous directions.

The term "Black-Board" is related to the chosen architecture that has been implemented to control the calibration on a distributed cluster. That is, many of the calculations are carried out on the local nodes of the compute cluster, and the global management (book keeping) is controlled through the use of databases. BBS can be operated in either Standalone mode, where a single sub-band is processed independently on a single compute node, or in Global Parameter Estimation mode, where multiple sub-bands are processed over distributed compute nodes. The latter is important when the signal-to-noise ratio of the data is low, and/or the full observed bandwidth of the data is needed to derive reliable calibration solutions.

The data workflow through BBS is shown in Fig. 5.8. First, BBS obtains the un-calibrated visibilities from the On-Line Application Processing (OLAP) system, which stores the observational data in the form of a Measurement Set. The calibration process is directed by a parameter set (parset) file supplied by the astronomer (the various parameters and suggested settings are discussed in the LOFAR Imaging Cookbook). The values for the various models used in the calibration (sky surface brightness distribution, instrument, ionosphere) are retrieved from the Parameter Database (parmdb), which is also updated during the calibration process. Using this information, BBS uses the self-calibration algorithm to solve the Measurement Equation and produce calibrated visibilities. Finally, the calibrated visibilities are then passed to the Imager, which carries out a Fourier transform to obtain an image

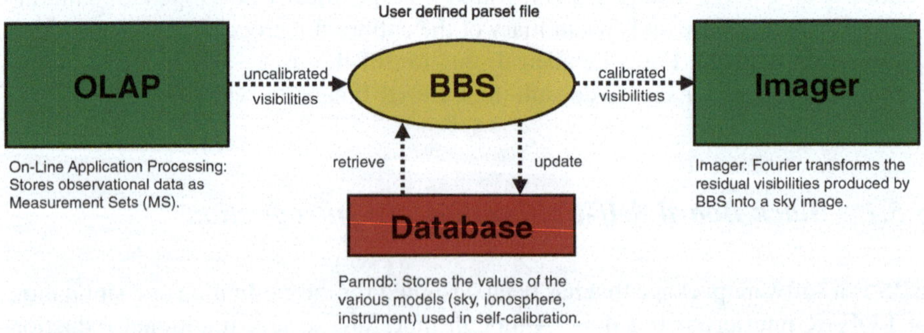

Fig. 5.8 The workflow of the BBS software package. Preprocessing (for example flagging and averaging) may take place within NDPPP, as described in Chap. 4, between the OLAP and BBS elements

of the sky surface brightness distribution and perform deconvolution. Note that BBS can also be used to subtract the sky model from the corrected data; in this case images of the residual visibilities can be made.

The BBS kernel can support a number of operations on a given Measurement Set, which are calculated locally on each compute node. They are as follows.

1. PREDICT: A set of model visibilities are simulated given a model for the sky surface brightness distribution, the instrument and the environment. Here, the model for the apparent sky model is produced by using the input sky surface brightness distribution and attenuating this by the appropriate model for the beam response. The Direct Fourier Transform (DFT) of this model is then taken for the regions of the Fourier space that are covered by the uv-sampling function.
2. SUBTRACT: The predicted contribution to the model visibilities of one or more sources are determined, and then subtracted from the observed (DATA) or calibrated (CORRECTED) visibilities to remove those sources from the data set.
3. CORRECT: The calibration tables are applied to the observed (DATA) visibilities to produce calibrated (CORRECTED) visibilities for a given reference (source) direction.
4. SHIFT: A phase shift is applied to the observed (DATA) visibilities so that they are relative to a new phase centre (position on the sky).
5. SOLVE: Calibration parameters/tables are calculated.

BBS forms a crucial part of the LOFAR imaging pipeline (see Chap. 8 for details).

5.5.2 Specifying the Sky Model

The sky model describes the surface brightness distribution of the sky. Within BBS, there are two ways to specify this. The first is by parameterising a sky image, typically found by using a de-convolved image of the initially calibrated data, which is produced using the CLEAN algorithm and its variants (Högbom 1974; Rau and Cornwell 2011), or by using images from publicly available low-frequency sky surveys, such as WENSS, VLSSr, TGSS-ADR1, GLEAM or MSSS (see above). This model is produced by fitting 2-dimensional elliptical Gaussian components or delta functions to the image, and parameterising the sky in terms of the flux-density, position, size, spectral index and polarisation of these multiple components. The model components are input to BBS via a text file, which can be easily manipulated by the astronomer. Here, patches of sources can be grouped together to determine the solutions toward a particular direction, which allows direction dependent effects to be solved for. This method is particularly useful if the dataset has many sources that contribute significantly to the sky brightness distribution and if direction dependent effects need to be solved for (which will be the case for the majority of LOFAR observations).

Alternatively, it is possible to use the de-convolved model that has been produced via the CLEAN algorithm, or more sophisticated methods such as compressed sensing (Garsden et al. 2015) or Bayesian inference (Junklewitz et al. 2016) to directly generate model visibilities via a Fourier transform. This method has the advantage of producing accurate models for bright and complex sources (for example, Cygnus A) that cannot be simply approximated by a set of Gaussian components, but has the disadvantage that only solutions in one direction can be derived.

Of course, it is possible to use a hybrid of both methods, depending on the nature of the data set. For example, a set of Gaussian model components can be used to characterise the wide-field sky brightness distribution and to determine the direction dependent effects. These components can then be subtracted from the Measurement Set, before imaging and self-calibration, using a complex model for the extended object of interest, is applied.

5.5.3 Calibration Strategies

Several strategies to calibrate the data from LOFAR have been developed during the initial commissioning phase that have been since refined and updated during the science phase. Again, the reader should consult the LOFAR Imaging Cookbook for additional details as here, only a summary is presented. There are three main calibration strategies that are used for the analysis of LOFAR data.

1. GAIN TRANSFER: The single direction independent amplitude and phase solutions for a strong calibrator field are used to calibrate the target field. This method is useful if there is no model for the target field sky brightness distribution, or if the target field has no bright sources in-beam that can be used to determine the initial phase and amplitude corrections. There are two ways to apply this method.
   - The observation cycles between a calibrator field, which is used to estimate the gain variations and transfer them to the target field, and the target field itself (called interleaved observations).
   - The observation uses the multi-beaming capability of LOFAR to simultaneously observe the calibrator field and target field (this helps to remove temporal variations).
2. GAIN CALIBRATION: (direction independent) Determines for a single direction the amplitude and phase solution and applies this to the visibility data. This method is useful when the target is a bright object in the centre of the field, or if the science goal can be achieved without reaching the thermal noise of the data.
3. GAIN CALIBRATION: (direction dependent) Solves for the amplitude and phase solution toward bright sources and subtracts them from the data, and then solves for the amplitude and phase solution toward the target. This method is useful for calibrating the data where the targets are faint objects in the field and is typically

required if the ionospheric conditions are particularly bad, if there is a dominant set of bright sources in the field, or if the science goals of the observation require a high level of calibration that reaches the thermal noise level.

Which of these methods the astronomer should apply is dependent on the way that the observation was carried out and on the science goal of the observation.

5.5.4 Inspecting and Flagging Solutions

In practice, calibration is a complex process and some care must be applied by the astronomer when deriving the calibration solutions. The quality of the calibration process can be determined by looking at the residuals between the model and corrected visibilities (although this requires a good signal-to-noise ratio to be effective) or by making an image to assess whether there is evidence for amplitude and phase errors. However, both of these methods require the calibration to be applied, and in the latter, for an image to be made, both of which can be time-consuming or compute-resource expensive. It is therefore advantageous to inspect the calibration solutions before applying them to the data.

Here, we also apply some a priori knowledge as to the expected variations of the amplitude and phase solutions as a function of time and frequency. Taking the amplitude solutions first, we would expect that they will be slowly varying, as the dominant source of change is due to the beam response of the individual stations. Note that source structure can still play a role, for example, an incomplete sky model coupled with bright sources entering the far side-lobes can cause unexpected variations in the amplitude solutions. An example of the amplitude solutions for a core and remote station have already been presented in Fig. 5.3. We see that the initial set of solutions for the remote station vary significantly over time, which given the slow varying nature of the beam response is not expected. Applying these solutions to the data would result in significant amplitude errors (in the case shown, the signal-to-noise ratio of the target was good enough for self-calibration to correct these errors, but this is not always the case).

The phase variations will change on a much shorter time-scale as these are dominated by the ionospheric fluctuations (see Fig. 5.9). These will be highly dependent on the observing conditions and so some care needs to be used when inspecting and interpreting phase solutions. However, it is generally the case that the solutions should change in a consistent way, that is, it should be possible to see a smooth variation in the phase solutions as a function of time.

For both the amplitude and phase corrections, any outliers or an appearance of random, noisy structure is evidence for poor quality solutions. These can be improved upon by increasing the signal-to-noise ratio of the data, by using a longer solution interval (see above) and recalculating the corrections, or by smoothing the solutions. However, in some cases, it may not be possible to make good corrections with either method, where it may be that there is bad data (due to RFI or un-modelled

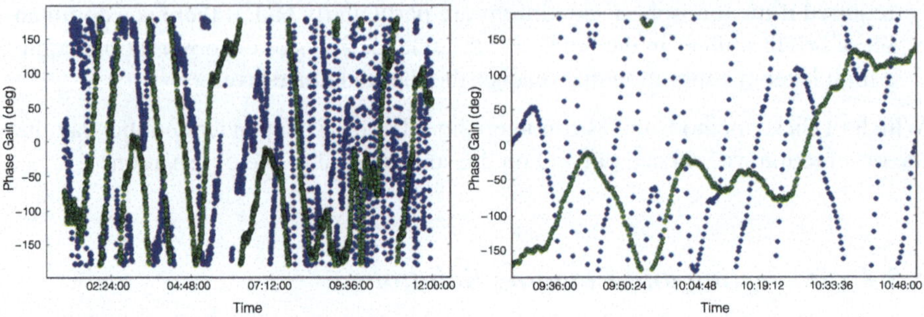

Fig. 5.9 Left: Example of the phase solutions, relative to a station in the superterp, obtained for a core (green; 3 km baseline) and remote (blue; 30 km baseline) station in the case of a 12-h observation of Cygnus A with a 4-s solution interval. The core station shows the expected, smoothly varying phase solutions for a station with short baselines, whereas the solutions for the remote station show significant small time-scale variability, as expected for long baselines. Right: The solutions over a $\sim$ 15 min period with fast phase variations. Although changing significantly, there is a clear trend in the solutions, with no major outliers or random structure, which suggests that the ionospheric corrections are good

signal). In these cases, it is best to flag those bad data or outlier solutions before applying the corrections to the observed data.

5.5.5 Direction Dependent Calibration

As has been discussed above, the wide field-of-view of LOFAR is sensitive to spatially varying ionospheric conditions, which will primarily result in direction-dependent phase solutions, and a spatially varying beam response, which will lead to direction-dependent amplitude solutions. Applying a single direction independent amplitude and phase solution to a LOFAR data set will remove the large-scale calibration errors in the data, and an image of the field can be made. However, the small-scale errors due to the locally varying ionospheric conditions and beam response will limit the quality of the image across the field. For example, one will often find a bright (>10 Jy) source somewhere within the imaged field. The side-lobes from such a source can limit the dynamic range as soon as the thermal noise is well below a mJy due to the varying ionospheric conditions or uncertainties in the beam response. This is typically the case for a 6-h synthesis with 60 MHz bandwidth in the HBA band. However, even without a bright source, there are many possible sources of errors.

To overcome these issues, solutions can be determined in several directions, which is referred to as direction dependent calibration. Similar to the direction independent calibration, this involves solving for the Measurement Equation, but instead of working in one direction it computes solutions in a number of directions

(s), such that the Measurement Equation becomes,

$$\mathbf{V}_{ij}^{\text{obs}} = \sum_s \mathscr{J}_{ij,s} \mathbf{V}_{ij,s}^{\text{model}}. \tag{5.11}$$

Direction dependent calibration can be specified within BBS by defining patches of sources, regions within which the spatially varying amplitude and phase solutions are considered to be constant; the size and hence number of patch regions are limited by the signal-to-noise ratio of the data, the ionospheric conditions and the compute resources available. Note that direction-dependent calibration solutions cannot be applied simultaneously to the visibilities; only one direction can be applied at once. The latter is the most significant issue since solving for multiple directions simultaneously is computationally expensive, even for the LOFAR dedicated processing cluster (CEP). It is for this reason that innovative calibration strategies and algorithms have been developed to calibrate for direction dependent effects within LOFAR data, which is the focus of the next section.

5.6 Advanced Calibration Techniques

Wide field observations, particularly at low frequencies have been carried out for decades with telescopes like the WSRT, VLA and GMRT, and many of the techniques that were developed for those telescopes have been modified and scaled-up to be used with LOFAR. For example, the peeling technique (Noordam 2004; Intema et al. 2009) and global calibration (Noordam and Smirnov 2010; Kazemi et al. 2013) have been widely used (see also the field based calibration method; Cotton et al. 2004). These two methods have evolved to become what is called facet calibration (Sect. 5.6.1), which divides the wide-field into smaller regions where the calibration is carried out, and using the whole field and sophisticated algorithms to solve for many directions simultaneously. We also note the increasing importance of advanced calibration strategies making use of Wirtinger derivatives (see Tasse 2014; Smirnov and Tasse 2015) and an associated imager (Tasse et al. 2018). All of these methods are also being tested with a view of applying them to the Square Kilometre Array (SKA).

5.6.1 A Facet Calibration Approach

The facet calibration method (van Weeren et al. 2016b), and its variants, is at the time of writing the standard approach used to calibrate wide-field datasets that show evidence for direction dependent effects, but require a high dynamic range. Here, after an initial direction independent calibration is carried out on the full data set, the model for the sky is subtracted from the uv-dataset, which in the case of perfect calibration across the field would leave just noise. However, there is typically un-modelled source structure remaining due to calibration errors. The sky is then divided into smaller regions, each typically centred on a bright source ($\sim$ 0.4 Jy) since they can provide the signal-to-noise ratio needed for calibration and are typically the sources that limit the dynamic range of the image. The sky model for each region around this bright source is added back into the uv-dataset and calibrated for separately. This can be considered as direction independent calibration over many small independent regions. The calibrated images (facets) for each region are then combined to form a mosaic of the field, where the calibration errors are now substantially reduced and the rms map noise approaches the expected thermal noise (van Weeren et al. 2016a).

An example of the improvement in image quality for four different directions within the same LOFAR data set is shown in Fig. 5.10. After the initial direction independent calibration, each of the bright objects show significant un-modelled side-lobe structure. After correcting for phase variations due to the ionosphere and residual clock off-sets between the stations, there is significant improvement in the maps in each direction, but there remain residuals. These are mainly amplitude errors due to the varying beam response; after correcting for these amplitude variations the side-lobe structure is now properly modelled, which shows that the calibration process has improved the imaging quality. Automating this facet based calibration approach is currently being implemented in the LOFAR imaging pipeline. However, the reader is strongly encouraged to refer to the LOFAR Imaging

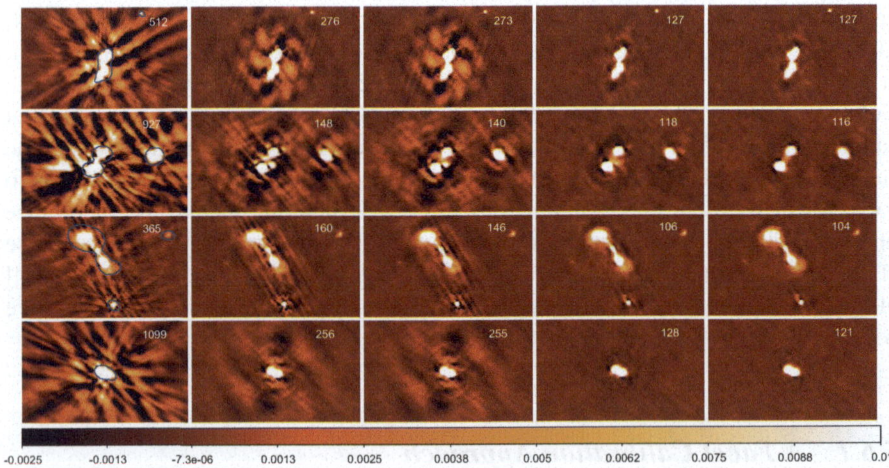

Fig. 5.10 An example of the improvement in calibration performance from using a facet-based approach (van Weeren et al. 2016b). For the four different directions (rows) the imaging results from a direction independent calibration (left), the results from solving for the TEC and clock phase variations in each direction (left-middle and middle) and the results from carrying out phase and amplitude self-calibration to additionally account for the beam response in each direction (right-middle and right). From van Weeren et al. ApJS (2016b) 223, 2, reproduced by permission of the AAS

Cookbook to understand the various steps in the process before applying it to their own data.

5.6.2 A Global Solve Using SAGECal

An alternative method to calibrate wide-field images for direction dependent effects is to solve for the whole field simultaneously, as is done using the Space Alternating Generalised Expectation Maximisation calibration technique (SAGECal; Kazemi et al. 2011). Similarly to the facet calibration approach, the field is divided into clusters of sources and then solved for in each direction to obtain the calibration solutions (Kazemi et al. 2013), but has the advantage that all of the sources in the field are taken into account at the same time. Although this should provide more accurate calibration solutions when compared to the facet calibration method, as the signal-to-noise ratio is higher through combining many more sources, it is also computationally more expensive. Therefore, the solutions are calculated using the expectation maximisation algorithm (Fessler and Hero 1994), which is more accurate and faster than traditional least squares minimisation techniques. From the early days of LOFAR operations, the SAGECal method produced wide-field images from LOFAR with the lowest rms noise levels and the highest dynamic range (Yatawatta et al. 2013), and also allowed measurements of large-scale polarisation (Jelić et al. 2015). An example of the improvement in the image quality over wide-fields of view is shown in Fig. 5.11.

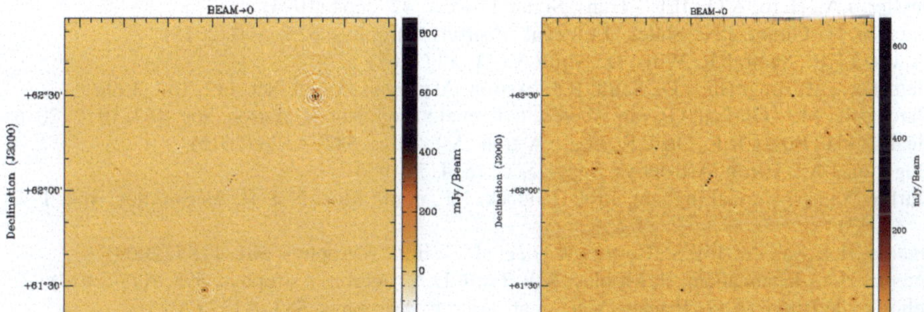

Fig. 5.11 An example of the improvement in calibration performance from using direction independent (left) and direction dependent (right) calibration techniques within SAGECal. The bright sources show significant side-lobe structure due to residual amplitude and phase errors, which limits the dynamic range for direction independent calibration. These side-lobes are significantly reduced when direction dependent solutions are obtained towards the problematic sources. This increases the dynamic range and allows the fainter radio source population in the field to be detected

5.7 Summary

In this chapter, we have presented a summary of the calibration challenges related to LOFAR data and their solutions, which should be seen as a continuing effort that aims to improve on the accuracy and efficiency of the calibration process. Many of these processes, including advanced calibration procedures discussed above are being implemented within the LOFAR Imaging Pipeline that will be run by the Radio Observatory. Of course, the level of calibration that is required to achieve the science goals for each experiment will be highly variable, but having a standardised process that calibrates each LOFAR dataset in a uniform way will be important for the legacy value of the telescope, and the accessibility of LOFAR to the wider astronomical community. However, as is the case for all data analyses, the input of the astronomer will be important to check the calibration reliability and image quality.

Acknowledgements We would like to thank Roberto Pizzo and George Heald for useful discussions, and the large number of developers and commissioners who have contributed to developing the calibration strategies that are summarised here.

References

Baars, J.W.M., Genzel, R., Pauliny-Toth, I.I.K., Witzel, A.: Astron. Astrophys. **61**, 99 (1977)
Cotton, W.D., Condon, J.J., Perley, R.A., et al.: In: Oschmann, J.M. Jr. (ed.) Ground-Based Telescopes. Proceedings of SPIE, vol. 5489, pp. 180–189. SPIE, Bellingham (2004)
Fessler, J.A., Hero, A.O.: IEEE Trans. Signal Process. **42**, 2664 (1994)
Garsden, H., Girard, J.N., Starck, J.L., et al.: Astron. Astrophys. **575**, A90 (2015)
Gower, J.F.R., Scott, P.F., Wills, D.: MmRAS **71**, 49 (1967)
Hamaker, J.P., Bregman, J.D., Sault, R.J.: Astron. Astrophys. Suppl. Ser. **117**, 137 (1996)
Hardcastle, M.J., Gürkan, G., van Weeren, R.J., et al.: Mon. Not. R. Astron. Soc. **462**, 1910 (2016)
Heald, G.H., Pizzo, R.F., Orrú, E., et al.: Astron. Astrophys. **582**, A123 (2015)
Högbom, J.A.: Astron. Astrophys. Suppl. Ser., **15**, 417 (1974)
Hurley-Walker, N., Callingham, J.R., Hancock, P.J., et al.: Mon. Not. R. Astron. Soc. **464**, 1146 (2017)
Intema, H.T., van der Tol, S., Cotton, W.D., et al.: Astron. Astrophys. **501**, 1185 (2009)
Intema, H.T., Jagannathan, P., Mooley, K.P., Frail, D.A.: Astron. Astrophys. **598**, A78 (2017)
Jelić, V., de Bruyn, A.G., Pandey, V.N., et al.: Astron. Astrophys. **583**, A137 (2015)
Jones, R.C.: J. Opt. Soc. Am. (1917–1983) **31**, 488 (1941)
Junklewitz, H., Bell, M.R., Selig, M., Enßlin, T.A.: Astron. Astrophys. **586**, A76 (2016)
Kazemi, S., Yatawatta, S., Zaroubi, S., et al.: Mon. Not. R. Astron. Soc. **414**, 1656 (2011)
Kazemi, S., Yatawatta, S., Zaroubi, S.: Mon. Not. R. Astron. Soc. **430**, 1457 (2013)
Laing, R.A., Riley, J.M., Longair, M.S.: Mon. Not. R. Astron. Soc. **204**, 151 (1983)
Lane, W.M., Cotton, W.D., van Velzen, S., et al.: Mon. Not. R. Astron. Soc. **440**, 327 (2014)
Macquart, J.-P., de Bruyn, A.G.: Mon. Not. R. Astron. Soc. **380**, L20 (2007)
Noordam, J.E.: In: Oschmann, J.M. Jr. (ed.) Ground-Based Telescopes. Proceedings of SPIE, vol. 5489, pp. 817–825. SPIE, Bellingham (2004)
Noordam, J.E., Smirnov, O.M.: Astron. Astrophys. **524**, A61 (2010)
Offringa, A.R., de Bruyn, A.G., Zaroubi, S.: Mon. Not. R. Astron. Soc. **422**, 563 (2012)

Offringa, A.R., McKinley, B., Hurley-Walker, N., et al.: Mon. Not. R. Astron. Soc. **444**, 606 (2014)
Pandey, V.N., van Zwieten, J.E., de Bruyn, A.G., Nijboer, R.: In: Saikia, D.J., Green, D.A., Gupta, Y., Venturi, T. (eds) The Low-Frequency Radio Universe. Astronomical Society of the Pacific Conference Series, vol. 407, p. 384. Astronomical Society of the Pacific, San Francisco (2009)
Perley, R.A., Butler, B.J.: Astrophys. J. Suppl. Ser. **230**, 7 (2017)
Pilkington, J.D.H., Scott, J.F.: MmRAS **69**, 183 (1965)
Rau, U., Cornwell, T.J.: Astron. Astrophys. **532**, A71 (2011)
Rengelink, R.B., Tang, Y., de Bruyn, A.G., et al.: Astron. Astrophys. Suppl. Ser. **124**, 259 (1997)
Roger, R.S., Bridle, A.H., Costain, C.H.: Astron. J. **78**, 1030 (1973)
Scaife, A.M.M., Heald, G.H.: Mon. Not. R. Astron. Soc. **423**, L30 (2012)
Smirnov, O.M.: Astron. Astrophys. **527**, A106 (2011)
Smirnov, O.M., Tasse, C.: Mon. Not. R. Astron. Soc. **449**, 2668 (2015)
Tasse, C.: Astron. Astrophys. **566**, A127 (2014)
Tasse, C., van der Tol, S., van Zwieten, J., van Diepen, G., Bhatnagar, S.: Astron. Astrophys. **553**, A105 (2013)
Tasse, C., Hugo, B., Mirmont, M., et al.: Astron. Astrophys. **611**, A87 (2018)
van der Tol, S., Jeffs, B.D., van der Veen, A.-J.: IEEE Trans. Signal Process. **55**, 4497 (2007)
van Weeren, R.J., Brunetti, G., Brüggen, M., et al.: Astrophys. J. **818**, 204 (2016a)
van Weeren, R.J., Williams, W.L., Hardcastle, M.J., et al.: Astrophys. J. Suppl. Ser. **223**, 2 (2016b)
Vedantham, H.K., Koopmans, L.V.E.: Mon. Not. R. Astron. Soc. **453**, 925 (2015)
Yatawatta, S., de Bruyn, A.G., Brentjens, M.A., et al.: Astron. Astrophys. **550**, A136 (2013)

Chapter 6
Error Recognition in LOFAR Data

Elizabeth Mahony

Abstract Identifying and removing errors in both the u,v plane and the image plane is essential to achieve the highest quality image. In this lecture, I describe how to recognise and remove some of the most common imaging artifacts. In particular, issues associated with wide-field imaging are discussed.

Being able to recognise and remove image defects is vital in achieving high quality images. In this chapter I briefly discuss some of the common imaging artifacts that occur in synthesis imaging, with a particular focus on issues that become more apparent when imaging the wide-fields observed by LOFAR.

6.1 Diagnosing Errors in the u,v Plane Compared to the Image Plane

Most of the image defects that we see are caused by bad data or errors in the measurement (u,v) plane. However, these errors also follow the Fourier transform relations introduced in Chap. 1 meaning that small, almost unrecognisable errors in the u,v plane can become quite obvious when transformed into the image plane. Likewise, significant errors in the u,v plane may be virtually impossible to detect in the image plane (besides perhaps a slight increase in the noise level). Of course it is still important to flag obviously bad data in the u,v plane, but this alone does not always guarantee the resulting image will be artifact-free. As the ultimate goal is to obtain a high quality image, identifying errors in the image plane can often be more useful.

E. Mahony (✉)
ASTRON, Dwingeloo, The Netherlands

Sydney Institute for Astronomy, School of Physics A2, The University of Sydney, Camperdown, NSW, Australia
e-mail: elizabeth.mahony@sydney.edu.au

G. Heald et al. (eds.), *Low Frequency Radio Astronomy and the LOFAR Observatory*, Astrophysics and Space Science Library 426,
https://doi.org/10.1007/978-3-319-23434-2_6

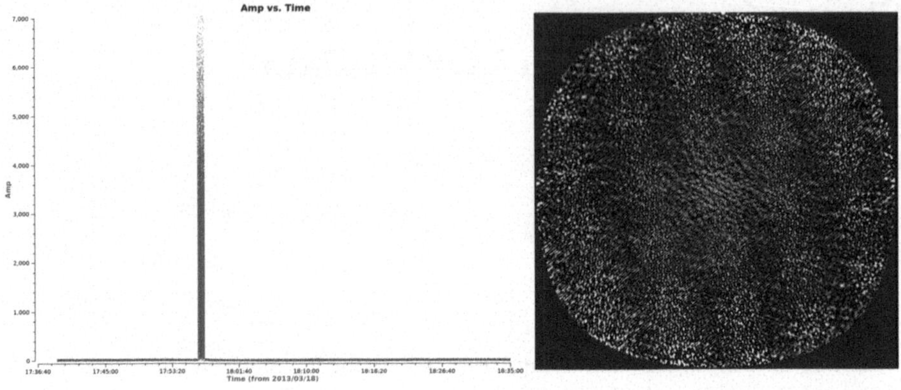

Fig. 6.1 An example of amplitude spikes manifesting as sinusoidal ripples once Fourier-transformed into the image plane

6.2 Recognising Calibration Errors in the Image Plane

We can use our knowledge of Fourier transform pairs to our advantage when trying to diagnose artifacts in the image. Searching for regular patterns or symmetries can provide hints to the origin of the defect. For example, amplitude errors will manifest as symmetric errors around sources, whereas errors in phase produce asymmetric artifacts. The Fourier transform of a sharp peak in amplitude in the visibility plane (such as RFI) produces a sinusoidal ripple throughout the image as shown in Fig. 6.1. A more thorough discussion of general error recognition in synthesis imaging is given in Ekers (1999).

6.3 Recognising Imaging Errors

Errors can occur not only due to bad data in the u,v plane, but also due to deconvolution errors. Negative bowls around bright sources are usually an indication that short spacings are missing, but can also indicate that the data has not been CLEANed to sufficient depth. Likewise, if there are artifacts around the sources that look very similar to the dirty beam (or point-spread function) this is a good indication that further CLEANing is needed.

The deconvolution algorithm chosen can also affect the image quality. For example, when interested in imaging extended, diffuse emission, the multi-scale clean algorithm (Cornwell 2008) can better recover large scale flux components. In addition to this, when imaging over large bandwidths it is important to take into

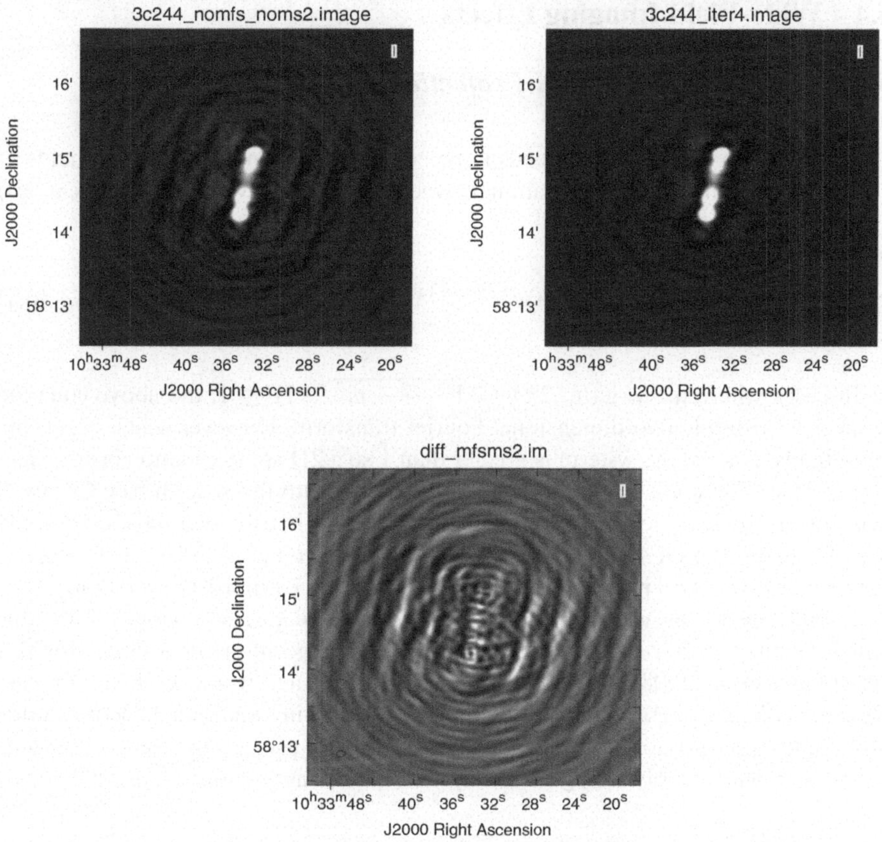

Fig. 6.2 The top panel shows LOFAR observations of 3C244.1. The image on the left used the standard CLEAN algorithm and the image on the right used multiscale-mfs CLEAN. The image below shows the difference between the top two images

account the spectral index of the sources using multi-frequency synthesis (mfs).[1] Figure 6.2 shows the differences in using multiscale-mfs CLEAN compared to the standard CLEAN algorithm.

[1] Technically multi-frequency synthesis simply refers to the technique of mapping the spectral channels to their correct frequencies to increase u,v coverage (Conway et al. 1990; Sault and Wieringa 1994). However, with the significantly increased bandwidths of many new or upgraded interferometers it is now possible to obtain an accurate spectral index across the band by modeling both the spatial and spectral structure simultaneously using the MS-MFS CLEAN algorithm (Rau and Cornwell 2011).

6.4 Wide-Field Imaging Effects

6.4.1 w-Projection and A-Projection Effects

One of the primary issues that needs to be considered when imaging wide fields of view is the w-term. Recall the relation between the observed visibilities ($V(u, v, w)$) and the sky brightness ($I(l, m)$):

$$V(u, v, w) = \int\int I(l, m) e^{-2\pi i[ul+vm+w(\sqrt{1-l^2-m^2}-1)]} dl dm \quad (6.1)$$

In cases where the w-term: $2\pi w(\sqrt{1-l^2-m^2}-1) \leq 1$, the above equation reduces to a simple two-dimensional Fourier transform. However, when observing large fields of view, the w-term is greater than 1 so a 2D approximation is no longer appropriate. There are several methods for dealing with the w-term (see Cornwell and Perley 1992; Perley 1999; Offringa et al. 2014), but the one most commonly used for imaging LOFAR data is w-projection (Cornwell et al. 2008). This projects the data onto the w=0 plane to then utilise the 2D Fourier transform relation.

In addition to the w-term, the LOFAR primary beam also changes with time and frequency, requiring the use of A-projection algorithms to account for this (Bhatnagar et al. 2008, 2013). Both these are combined in the LOFAR imaging software AWImager (see Chap. 8). Imaging the data using traditional deconvolution algorithms (that do not account for the w-term or A-projection effects) will lead to artifacts, particularly in regions away from the phase centre. Figure 6.3 shows

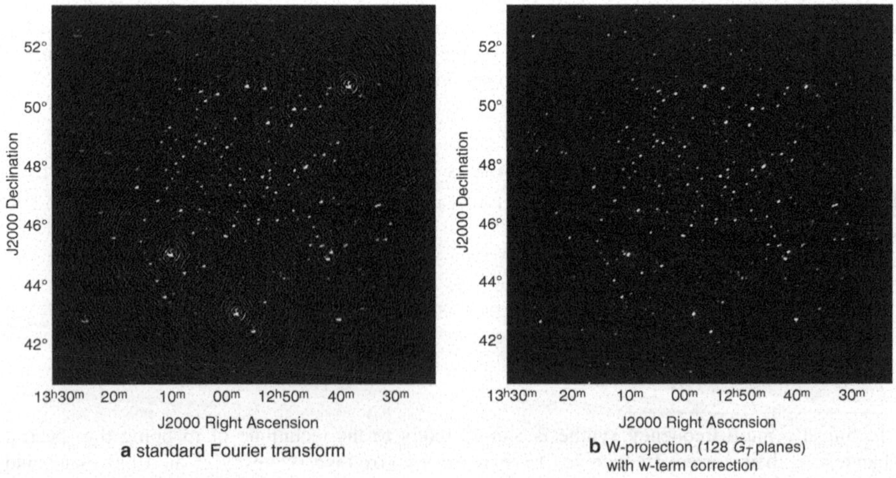

Fig. 6.3 Images obtained using the standard 2D Fourier transform (**a**) and correcting for the w-term using the WSclean algorithm (**b**). Images: A. Offringa

the difference in the image quality obtained by imaging wide-field data using the standard 2D Fourier transform compared to applying w-correction. Different algorithms used to correct for the w-term, including AWimager, are described in more detail in Chap. 8.

6.4.2 Bandwidth-Smearing and Time-Average Smearing

The large bandwidth of LOFAR can also produce imaging artifacts if not handled carefully. When making an image, the visibilities are gridded as if they were monochromatic, i.e. the width of the band is not accounted for. This means that if the data is averaged too much in frequency, sources will appeared smeared in the radial direction and the peak flux density will have decreased (although the integrated flux should remain the same). This effect gets progressively worse the further away from the phase centre the source is.

Equation (6.2) provides a guideline to what frequency resolution is required to avoid bandwidth smearing at different resolutions (assuming a square bandpass and gaussian beam). Here $\frac{I}{I_0}$ refers to the reduction in peak flux due to smearing, $\Delta\nu$ is the channel width, ν_c is the central frequency, r is the distance from the phase centre and θ is the resolution.

$$\frac{I}{I_0} = \frac{\sqrt{\pi}}{2\sqrt{\ln 2}} \frac{\theta \nu_c}{r \Delta\nu} \operatorname{erf}\left(\sqrt{\ln 2} \frac{r \Delta\nu}{\theta \nu_c}\right) \tag{6.2}$$

See also the related discussion in Chap. 11. A similar effect occurs when data is averaged in time, but in this case the sources are smeared in the tangential direction. While both of these effects are properties of radio interferometry in general, the dependence on distance from the pointing centre means that more care has to be taken when imaging large fields of view. This is why low-frequency arrays need high resolution in both frequency and time. For the detailed mathematics of bandwidth and time-averaging smearing the reader is referred to Bridle and Schwab (1999).

6.4.3 Sidelobes from Bright Sources in the Field

The large field of view of LOFAR means that it is highly likely that there will be at least one bright source ($S > 10$ Jy) in or close to the field. Depending on the flux density of the source, this can produce sidelobes throughout the image. Peeling is usually required to effectively remove these sources in order to increase the dynamic range and image fainter sources. Figure 6.4 (top image) shows the Lockman Hole field as observed with LOFAR. The bright radio source 3C244.1 (~35 Jy at 150 MHz) is approximately 2° away from the phase centre, but the sidelobes from this source contaminate the entire field of view.

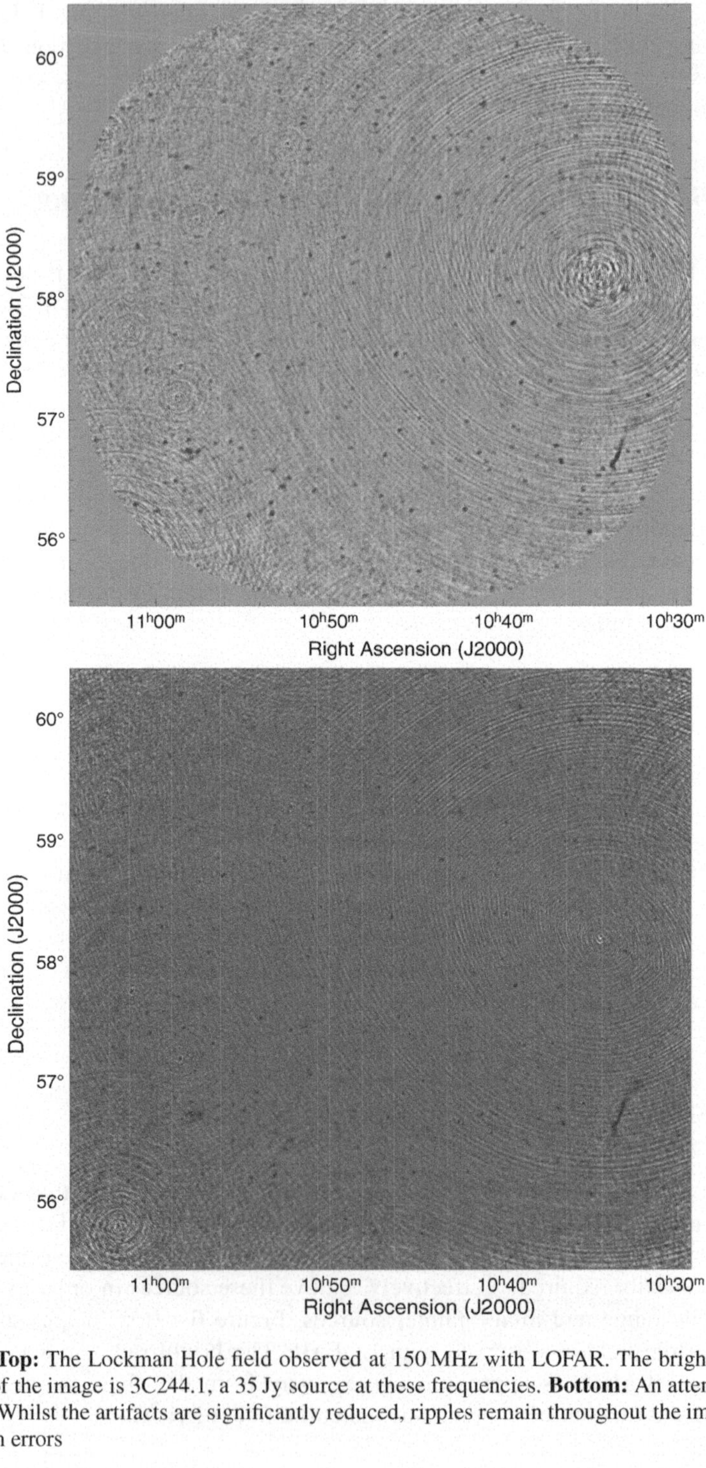

Fig. 6.4 **Top:** The Lockman Hole field observed at 150 MHz with LOFAR. The bright source to the west of the image is 3C244.1, a 35 Jy source at these frequencies. **Bottom:** An attempt to peel 3C244.1. Whilst the artifacts are significantly reduced, ripples remain throughout the image due to calibration errors

In order to peel sources properly, we need to be able to calibrate the data in that direction. Figure 6.4 (bottom image) shows an example of the artifacts that remain after 3C244.1 has been subtracted in the u,v plane, but not calibrated well enough. While the ripples are significantly reduced, they still remain at the level of 5–10 mJy/beam in this image. The residual flux of 3C244.1 is less than 1% of the total flux, but if not calibrated correctly this will still leave artifacts in the image. In this particular case, the artifacts were due to delay errors on the longer baselines. This dataset was imaged using 2 MHz bandwidth, but the delay errors cause phase decorrelation within that bandwidth. Remember that only the LOFAR core stations are on a single clock!

Figure 6.5 shows the result after successfully peeling 3C244.1. As often happens when identifying and removing errors, it has now become clear that other sources just outside the primary beam also need to be peeled in this image. Identifying and removing errors is an iterative process that requires constant re-evaluation after each step.

6.4.4 Direction Dependent Effects

One of the biggest difficulties in observing at low frequencies is correcting for ionospheric distortions. In images, these effects are observed as phase errors which vary significantly with time and viewing direction. LOFAR's large field of view means that sources across the image are being observed through different parts of the ionosphere and therefore require different phase corrections. This process of performing direction-dependent calibration is very computationally expensive and, at the present time, is one of the biggest limitations in achieving good quality LOFAR images at high resolution. Figure 6.6 shows the artifacts that are visible without doing any direction-dependent calibration (top) and after applying direction-dependent calibration (bottom), for example as described in Chap. 5.

6.5 Can You Do Science with It?

On a final note, it is important to keep in mind the scientific value of the image. You can produce an excellent image which is free of any obvious artifacts, but if the flux scale is wrong or the source positions are incorrect it is virtually useless. It is important to always crossmatch your image with other catalogues to check the flux densities and positions. For LOFAR HBA observations the 7C catalogue provides a good reference at 151 MHz (Riley et al. 1999), as does the recent TIFR GMRT Sky Survey Alternative Data Release (TGSS-ADR1; Intema et al. 2017); for LBA the VLSSr catalogue at 74 MHz is an excellent reference point (Lane et al. 2014).

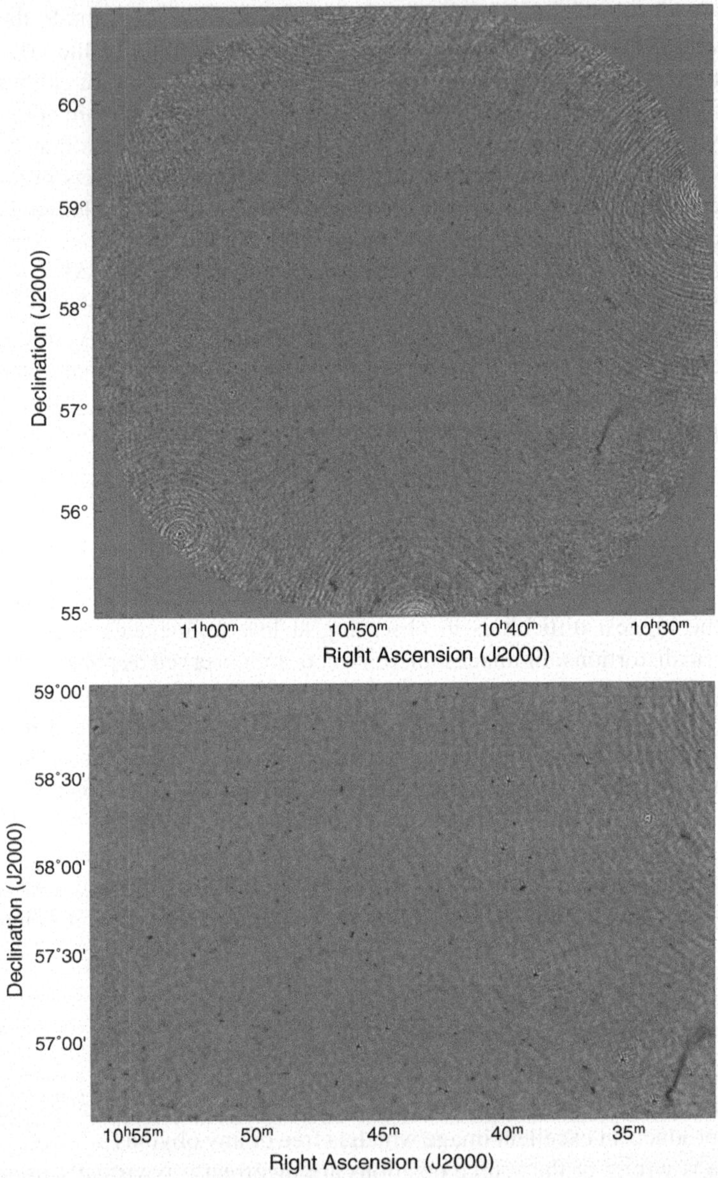

Fig. 6.5 **Top:** The Lockman Hole field after peeling 3C244.1. Other sources at the end of the field of view also need to be peeled. **Bottom:** A zoom-in of the central regions of the Lockman Hole field. Faint artifacts are still noticeable around 3C244.1 (top, righthand side of the image), but the ripples no longer extend throughout the full image

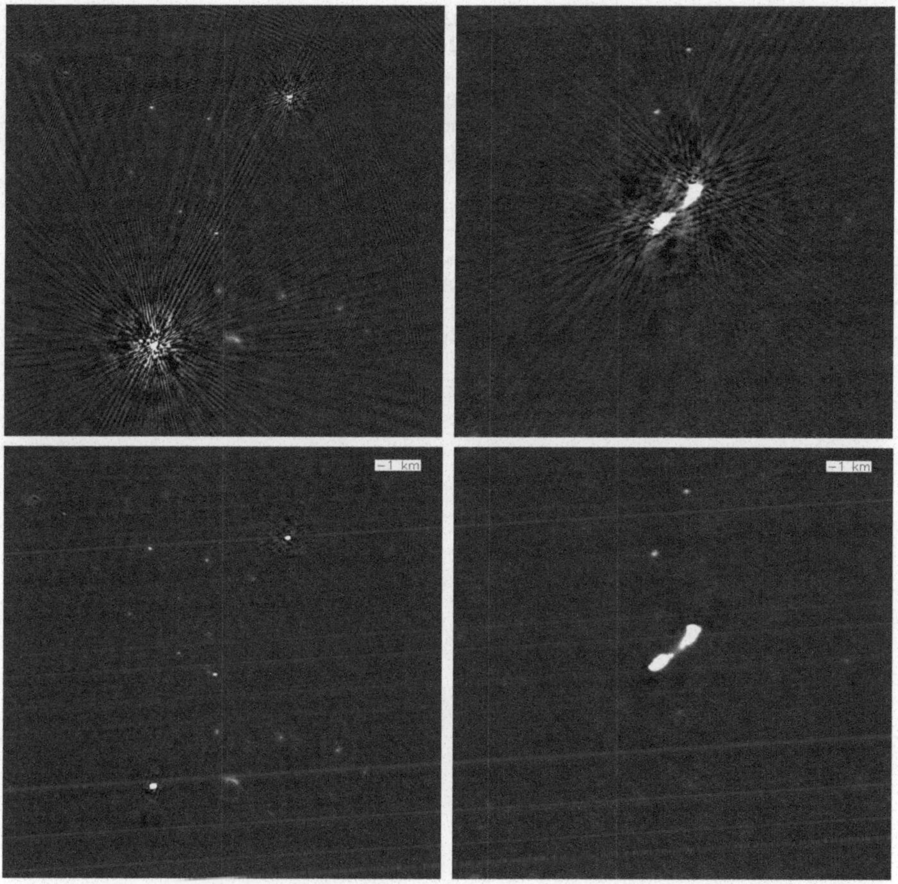

Fig. 6.6 **Top:** LOFAR HBA image of the Bootes field after direction-independent calibration and self-calibration. **Bottom:** The same field after direction-dependent calibration. Image credit: Wendy Williams

Acknowledgements This chapter has drawn heavily from numerous lectures given at previous synthesis imaging schools available online as well as the VLA white book (Synthesis Imaging in Radio Astronomy, editors: Taylor, Carilli & Perley).

References

Bhatnagar, S., Cornwell, T.J., Golap, K., Uson, J.M.: Astron. Astrophys. **487**, 419 (2008)

Bhatnagar, S., Rau, U., Golap, K.: Astrophys. J. **770**, 91 (2013)

Bridle, A.H., Schwab, F.R.: In: Taylor, G.B., Carilli, C.L., Perley, R.A. (eds.) Synthesis Imaging in Radio Astronomy II. Astronomical Society of the Pacific Conference Series, vol. 180, p. 371 (1999)

Conway, J.E., Cornwell, T.J., Wilkinson, P.N.: Mon. Not. R. Astron. Soc. **246**, 490 (1990)

Cornwell, T.J.: IEEE J. Sel. Top. Signal Process. **2**, 793 (2008)
Cornwell, T.J., Perley, R.A.: Astron. Astrophys. **261**, 353 (1992)
Cornwell, T.J., Golap, K., Bhatnagar, S.: IEEE J. Sel. Top. Signal Process. **2**, 647 (2008)
Ekers, R.D.: In: Taylor, G.B., Carilli, C.L., Perley, R.A. (eds.) Synthesis Imaging in Radio Astronomy II. Astronomical Society of the Pacific Conference Series, vol. 180, p. 321 (1999)
Intema, H.T., Jagannathan, P., Mooley, K.P., Frail, D.A.: Astron. Astrophys. **598**, A78 (2017)
Lane, W.M., Cotton, W.D., van Velzen, S., et al.: Mon. Not. R. Astron. Soc. **440**, 327 (2014)
Offringa, A.R., McKinley, B., Hurley-Walker, N., et al.: Mon. Not. R. Astron. Soc. **444**, 606 (2014)
Perley, R.A.: In: Taylor, G.B., Carilli, C.L., Perley, R.A. (eds.) Synthesis Imaging in Radio Astronomy II. Astronomical Society of the Pacific Conference Series, vol. 180, p. 383 (1999)
Rau, U., Cornwell, T.J.: Astron. Astrophys. **532**, A71 (2011)
Riley, J.M.W., Waldram, E.M., Riley, J.M.: Mon. Not. R. Astron. Soc. **306**, 31 (1999)
Sault, R.J., Wieringa, M.H.: Astron. Astrophys. Suppl. Ser. **108**, 585 (1994)

Chapter 7
Ionospheric Effects

Maaijke Mevius

Abstract The ionosphere is a partly ionized turbulent plasma layer in the upper atmosphere, with an electron density that is highly variable in space and time. An electromagnetic signal passing through this layer experiences a wavelength dependent delay. This dispersive delay is a major issue for calibration of low frequency radio interferometric data. The time variability requires high time resolution phase solutions, whereas the spatial variation forces direction dependent calibration. In the presence of the Earth's magnetic field, the delay is slightly different for right and left circular polarization, leading to a rotation of the linear polarization angle, an effect which is well known as Faraday rotation. Faraday rotation is an issue that needs to be addressed when studying polarization, but its differential effect can also be observed for unpolarized signals in the cross correlation products at low frequencies or long baselines. In this chapter we will discuss the above mentioned effects of the ionosphere on LOFAR data as well as methods to correct for them.

7.1 Introduction

The ionosphere is an ionized layer in the upper atmosphere, ranging from about 50–1000 km above sea level. Ionization happens through solar irradiation, while at night recombination causes the ionization level to decrease. The molecule density decreases with increasing height, which combined with the radiation depth leads to a specific ionized density profile which has its maximum around 300–400 km. Traditionally the ionosphere is divided into multiple layers. The layer with the maximum density is dubbed the *F-layer*, and it is the only significantly ionized layer during night time. During the day, the lower *D-* and *E-layers* (between 50 and 150 km) become more heavily ionized. Often, for simplified analyses an assumption is made in which the ionosphere is described by an infinitely thin layer at a fixed

M. Mevius (✉)
ASTRON, Dwingeloo, The Netherlands
e-mail: mevius@astron.nl

G. Heald et al. (eds.), *Low Frequency Radio Astronomy and the LOFAR Observatory*, Astrophysics and Space Science Library 426,
https://doi.org/10.1007/978-3-319-23434-2_7

altitude. This *thin screen* method appears to be a helpful simplification in many cases.

For the analysis of electromagnetic propagation through the ionosphere, only the density of ions is important. The thickness of the ionosphere is measured in units of total electron content (TECU), which is the integrated electron density along the line of sight in units (1 TECU $= 10^{16}e^-/\text{m}^2$). Typical values for the total electron content above LOFAR vary from 1 TECU (nighttime, winter) to 50 TECU (daytime, summer).

When a signal travels through the ionosphere at a non-zero zenith angle, one should take into account the excess pathlength through the ionosphere due to the angle. One distinguishes vertical TEC (zenith angle = 0) from slant TEC. The correction factor, assuming a single layer at fixed height is:

$$1/\cos\left[\arcsin\left(\sin(\alpha)\cdot\frac{R_e+h_{ant}}{R_e+h_{ion}}\right)\right], \tag{7.1}$$

where α is the zenith angle, R_e is the Earth radius, and h_{ant} and h_{ion} are the altitude of the antenna and the assumed ionospheric layer, respectively.

Historically, measurements of the structure of the ionosphere were performed using a special radar for the ionosphere, an *ionosonde*, which makes use of the reflective properties of the ionosphere. The higher the electron density the higher the frequency of a reflected wave, and thus by transmitting a range of frequencies one can measure the density profile of the ionosphere above the ionosonde. Nowadays, satellites and GPS receivers on Earth continuously monitor the ionosphere. The GPS data of many receivers is available online at several ftp-servers, shortly after the data has been taken. Also, several groups (e.g. Center for Orbit Determination in Europe[1,2]) fit a thin layer coarse ionospheric model to the GPS data and also publish these data in IONEX format (Schaer and Gurtner 1998).

The ionospheric electron density has a complex and highly dynamic structure. The diurnal pattern due to daytime solar irradiation and recombination at night is just one of the examples of the time variation of the ionosphere. Variation in solar activity, lightning, earthquakes and other effects, all can cause time varying density fluctuations. As can be imagined, the largest gradients in electron density appear during sunrise and sunset.

It was found that the spatial structure of the ionospheric electron density is turbulent and can best described by a Kolmogorov structure function:

$$\langle\phi_{ij}\rangle = (S_{ij}/S_0)^{\beta}, \tag{7.2}$$

where $\langle\phi_{ij}\rangle$ is the phase variance between two points, S is the distance between the points, and S_0 is the diffractive scale at which the phase variance is 1 rad. For

[1]ftp://ftp.aiub.unibe.ch/CODE/.

[2]ftp://gnss.oma.be/gnss/products/IONEX/.

pure Kolmogorov turbulence the power-law index β is $5/3$. The typical diffractive scale at mid latitudes at 150 MHz varies between 2 and 40 km. If it gets close to the Fresnel scale which is ~1 km for LOFAR stations, one can get amplitude decorrelation at stations. This condition is called scintillation and usually prevents normal observations.

On top of this turbulent structure, traveling waves are often observed, with amplitudes larger than those of the Kolmogorov turbulence. These are named traveling ionospheric disturbances (TIDs). The exact mechanism causing these TIDs is not well understood. The typical timescale of these waves is on the order of 10–20 min, although sometimes much shorter waves are observed. Variations are also observed on large scales aligned with geomagnetic field lines (Loi et al. 2015).

The temporally- and spatially-varying nature of the ionosphere puts limits on the calibratability of its effects on radio signals. These effects will be discussed in the next sections.

7.1.1 Electromagnetic Propagation

The main effect of the ionized plasma on the electromagnetic signal passing through it is a frequency dependent delay of the signal. Since an interferometer is only sensitive to phase differences, only the TEC difference between the two signal paths to two stations making up an interferometric baseline results in a phase error due to the dispersive delay. Using phases from gain calibration one can measure the relative TEC above the stations. The structure of the ionosphere is such that the phase error is time variable and increases with baseline length. Dispersive delay and the correction for it will be discussed in Sect. 7.2.

A second effect on the signal, discussed in Sect. 7.3, is Faraday rotation. This is the rotation of the linear polarization angle that occurs if a signal travels through a plasma in the presence of a magnetic field, in this case the Earth's magnetic field. Unlike the dispersive delay, the (time variable) rotation of the polarization angle of a polarized signal is sensitive to the absolute TEC. However, as will be discussed below in Sect. 7.3.1, for longer baselines and lower frequencies, the difference in rotation angle also becomes visible in the cross product of the two signal streams of an interferometer even if the signal is not polarized. This is called *differential Faraday rotation*.

The main effects mentioned above will be explored in the next sections. Second order effects include differential refraction, which is in fact a geometrical effect due to refraction and the curvature of the Earth, and results in a relative shift of the position of sources in the field of view. The effect is sensitive to absolute TEC values. First order corrections to the data could be applied using GPS TEC values, although this correction is currently not available in the standard LOFAR software. It will not be discussed in more detail here, but a good treatment of this topic can be found in Thompson et al. (2001).

7.2 Dispersive Delay

The ionized plasma has a refraction index n_p that differs from unity by the signal frequency (ν) dependent relation:

$$n_p = \sqrt{1 - \left(\frac{\nu_p}{\nu}\right)^2} \approx 1 - \frac{1}{2}\left(\frac{\nu_p}{\nu}\right)^2 = 1 - \frac{40.3}{\nu^2} \cdot N_e, \tag{7.3}$$

where ν_p is the plasma frequency, which depends on the electron density N_e. The approximation is only valid in the limit $\nu \gg \nu_p$. Since the plasma frequency is on the order of 10 MHz, for the lowest LOFAR frequencies higher order terms should also be taken into account. In the following we assume the signal frequency is large enough that Eq. (7.3) is valid.

As a consequence of the aberrant refractive index of the ionized plasma, the electromagnetic waves penetrating the plasma are subject to a frequency dependent delay, depending on the integrated electron density along the line of sight:

$$\text{ionospheric excess path length} = 40.3/\nu^2 \cdot \int N_e \, dl. \tag{7.4}$$

This results in a phase error of:

$$\phi = 8.4479745 \cdot 10^9 \cdot \text{TEC}/\nu, \tag{7.5}$$

with TEC expressed in TECU. Since the paths of the waves are different for different stations, thereby piercing through a different part of the ionosphere, the net effect on the interferometric signal is a time and frequency dependent phase error, that only depends on the difference in integrated TEC along the line of sight. Furthermore, for a large field of view, the total electron content can vary from direction to direction within a single beam. For the longest Dutch LOFAR baselines the differential integrated TEC at night is in the order of 0.5–1 TECU. Within a single HBA beam, the typical variation is ~0.01 TECU. These differential TEC values can increase significantly during sunrise and sunset, and at periods when the ionosphere is much wilder (scintillation conditions). From the above numbers and Eq. (7.2) one can deduce that the typical phase error at the longest Dutch baseline results in 2–4 full phase rotations over the HBA-low bandwidth between 110 and 170 MHz. For LBA the situation is much worse: between 10 and 20 full 2π rotations from 30 to 80 MHz. Therefore, one needs to solve for individual phases at least per 195 kHz subband, or, if the S/N does not allow this, one can combine subbands but take into account the frequency dependent variation of the phases. The latter is possible in the LOFAR software by solving in BBS (Shulevski 2017) for TEC (direction independent) and a constant gain. Due to the non stationary character of the ionosphere, one should also solve with small time intervals. For HBA-low including remote stations, typically one phase solution per 10 s and 1 subband is sufficient.

A major problem of ionospheric calibration is the spatial variance of ionospheric phase errors. For a large field of view, one needs to correct different directions with their own phase corrections. It also has the consequence that the ionospheric phase errors are different for a calibrator and the field of interest, therefore one cannot simply transfer the phase solutions of the calibrator to the target field. And, even if one is only interested in a small field of view, bright sources outside the field of view need to be subtracted with their own phase corrections at the same resolution. Strategies to correct the direction dependent phases will be discussed in Sect. 7.2.2.

7.2.1 Separation of Phase Effects

As discussed in the previous section, the ionosphere forces self calibration of phases with high time and frequency resolution. The phases obtained this way do not only contain the ionospheric errors, but also other (e.g. instrumental) effects. All LOFAR core stations nowadays share the same clock resulting in tiny and stable clock errors, but for the remote stations this is not the case. Therefore, for remote stations, another significant contribution to the time variable phase errors are the drifting clock errors. A single clock difference $\Delta\tau$ results in phase error of:

$$\Delta\phi = 2\pi \cdot \nu \cdot \Delta\tau. \tag{7.6}$$

Apart from a constant ϕ_0 offset, these are the dominant contributions to phase errors in LOFAR.

Since the ionospheric and clock delays have a different frequency behavior, one can use the large bandwidth of LOFAR to separate the two effects. This can be useful for several reasons:

- Since the clock errors are only instrumental and not direction dependent, they can in principle be deduced for arbitrary directions from simultaneous calibrator phases. One can then reduce the phase errors in the field of interest by applying only the clock delays derived from the calibrator phases.
- The ionospheric phases give a very good indication of the ionospheric conditions during the observation. This can be used to discard times with heavy ionospheric fluctuations or to predict the remaining phase errors after self calibration in one or more directions.
- Finally, one can use the ionospheric phases to generate an *ionospheric phasescreen* over the field of view. This screen can be used to interpolate the phase solutions to directions where there are no bright sources to generate direction dependent solutions. This approach will be discussed in more detail in Sect. 7.2.2.

Separation of the clock and ionospheric solutions is done by fitting for A (delay values in s) and B (ΔTEC in TECU):

$$\Delta\phi(\nu) = A \cdot 2\pi\nu - B \cdot 8.45 \cdot 10^9/\nu \tag{7.7}$$

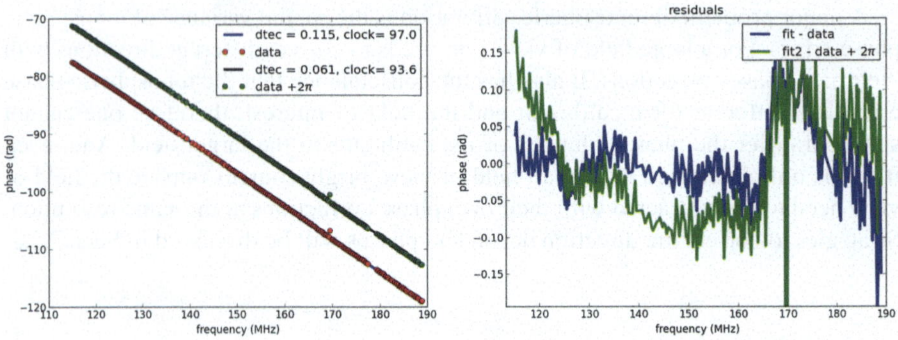

Fig. 7.1 Illustration of effect of 2π phase wraps in Clock/TEC separation on HBA calibration phases. Left:The lower line corresponds the best fit to the phases (red dots: unwrapped to match the fit), fitted parameters are given in the legend. The green dots and line correspond to a positive 2π shift of the data, the corresponding parameters have a shift in delay of about 3.5 ns and in TEC of 0.05 TECU. The residuals are shown on the right. The χ^2 of both fits only differ minorly

on the differential phase solutions over a large bandwidth. The bandwidth needs to be large because over small frequency ranges the frequency behavior of the second term can hardly be distinguished from a linear gradient. Typically for HBA-low observations it is recommended to use a 40 MHz bandwidth. For LBA a 20 MHz range should be sufficient.

A second complication is the problem of 2π phase ambiguities. If $\Delta\phi$ is a solution for the phase difference between 2 stations, so is $\Delta\phi + 2\pi$. This leads to local minima in χ^2 space, as illustrated in Fig. 7.1. The solutions corresponding to the local minima are separated by fixed steps in clock and ionospheric delay, which solely depends on the bandwidth used for the fit. One needs to initialize the parameters A, B with values close to their true solutions in order to obtain the right minimum. Since the values of both the clock and the TEC vary slowly in time, given sufficient time resolution of the phase solutions, the problem reduces to finding a good initialization for the first time slot and initializing the parameters of subsequent time slots with the solution of the previous one. With a good initialization of parameter A and B the phase solutions can be unwrapped along the frequency axis, in which case the fit of Eq. (7.7) to the data reduces to a linear fit.

In practice it is recommended to perform the fit in two iterations, removing possible 2π phasewraps by inspecting the time averaged residuals, or the spatial correlation between dTEC solutions of different stations. The latter is illustrated in Fig. 7.2, where the deviations of single stations to a linear gradient of the TEC solutions over the LOFAR array correspond to an offset of exactly one 2π phase wrap.

Clock/TEC separation is not yet available in the standard LOFAR pipeline, but an implementation exists in the Losoto package.[3]

[3]https://github.com/revoltek/losoto—see also Shulevski (2017).

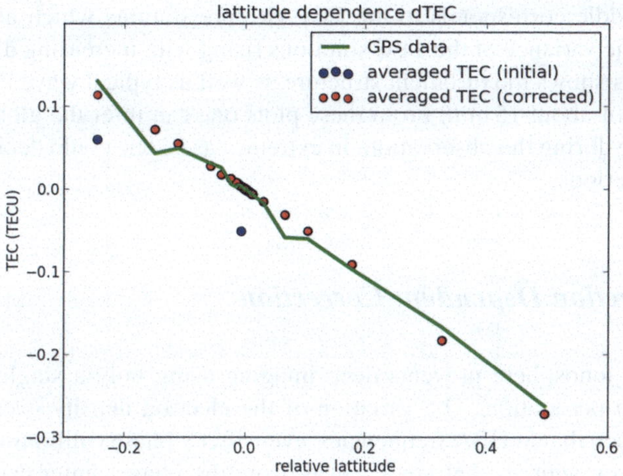

Fig. 7.2 Time averaged TEC solutions per station versus the relative latitude position of the station with respect to CS001HBA0. To first order the differential TEC values form a linear gradient over the array in the NS direction. Deviations of single station TEC solutions from this gradient correspond to a 2π phase error in the initialization. The green line shows the time averaged differential TEC values from GPS measurements

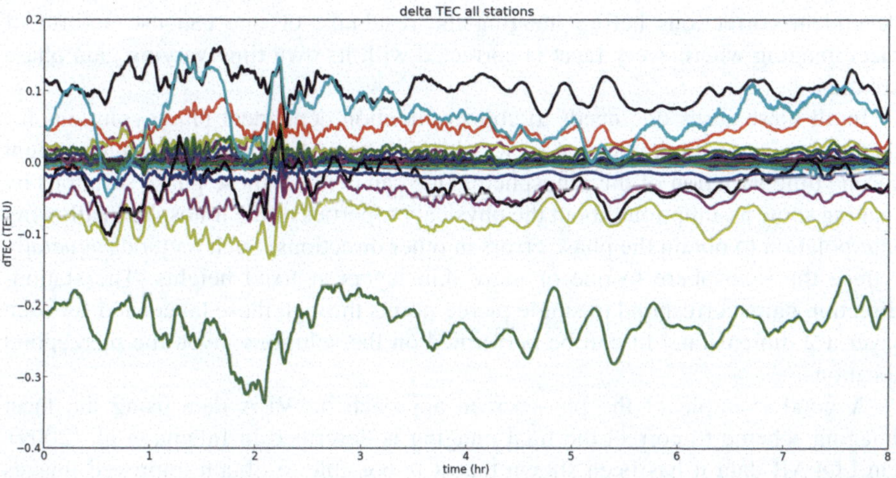

Fig. 7.3 Solutions of differential TEC with respect to CS001HBA0 versus time for HBA 3C196 data, after Clock/TEC separation on the calibration phases. Different lines correspond to different stations

The resulting TEC values after a successful Clock/TEC separation on HBA calibrator data are shown in Fig. 7.3. These are differential TEC solutions of all stations (with respect to the core station CS001HBA0) of a typical 8 h winter nighttime observation. Different lines correspond to different stations. The thick belt in the middle corresponds to the LOFAR core stations which all see similar TEC values, the variance of the TEC solutions rising with increasing distance to the core. The plots shows the turbulent structure as well as typical wave like structures with a period of about 15 min. From these plots one can infer the global quality of the ionosphere during the observation. In extreme cases, one could decide to discard certain wild periods.

7.2.2 *Direction Dependent Correction*

The turbulent ionosphere prevents deep imaging using only a single ionospheric phase solution per station. The variation of the electron density over the field of view can be such that at HBA frequencies, even after a phase calibration of a central bright calibrator, sources that are 1° away from the phase center can suffer from remaining phase errors of a full 2π rotation or more. To correct these effects one needs to perform a direction dependent calibration.

As discussed in Chaps. 5 and 8, direction dependent corrections cannot be directly applied to the visibilities. Using A-projection (Chap. 8) it is possible to apply a smooth interpolation to the direction dependent corrections during imaging. Another option is to subtract all known sources with their known direction-dependent corrections before imaging the residuals, or one can use a form of facet imaging where every facet is corrected with its own time varying gain phase solution.

In all these cases one needs to collect direction dependent phase solutions for the bright sources in the field of view, with high enough time resolution to account for the time variation of the ionosphere. One can connect these phase solutions by making some assumptions about the physical properties of the ionosphere, allowing interpolation to obtain the phase errors in other directions. Such methods generally reduce the ionosphere to one or more thin layers at fixed heights. The station-direction pairs correspond to single pierce-points through these layers and for each layer a 2 dimensional fit can be performed on the solutions given the piercepoint locations.

A good example of the phasescreen approach on VLA data using the facet imaging scheme to correct the final imaging is described in Intema et al. (2009). On LOFAR data it has been shown that it is possible to obtain improved images on the shorter baselines (<10 km) using a single screen and A-projection (Heald et al. 2015). A final strategy for dealing with the rapidly varying direction dependent phase effects has not been developed at the time of writing. It is likely that the large layout of full LOFAR does not allow the description of the ionosphere by a single screen and either a multiscreen approach or 3D tomography (e.g. Koopmans 2010) is required.

One of the issues in constructing an ionospheric phasescreen are other phase errors that enter the phase solutions. Even after successful Clock/TEC separation there are still remaining phase effects like beam errors. Especially if these errors are direction dependent and instrumental, as opposed to effects which are smooth or constant over the sky, this can complicate the interpolation of solutions of more stations in several directions. Furthermore, the relative positions of the pierce-points change with changing altitude of the layer. One should carefully chose the assumed height of the phasescreen. It is generally assumed at the position of the bulk ionosphere, between 200 and 400 km, varying between day and night.

As mentioned, a final strategy for dealing with ionospheric phase errors in LOFAR has not yet been developed. Currently, the most fruitful approach for HBA seems to be a deep peeling, making use of the fact that the amplitude (mostly beam) errors very only slowly in time, such that only the direction dependent phases need to be solved with high time resolution. The phase solutions obtained in this way can be combined in a separate phasescreen per facet to apply correction during imaging and deconvolution. In case the phasescreen approach is desired, A-projection is available in the standard LOFAR imager. In case of LBA data the number of peeling directions will be limited due to the limited number of sources in the field of view with high enough S/N.

7.3 Faraday Rotation

The combination of an ionized plasma and a magnetic field leads to an effect known as Faraday rotation. This rotation of the linear polarization angle is discussed further in Chap. 10. The linear polarization angle rotates by an amount (β) which is proportional to λ^2, and the constant of proportionality is called the rotation measure (RM). The RM is given by the integral of the product of the electron density n_e and the parallel magnetic field $B_{||}$ along the line of sight s:

$$\beta = \mathrm{RM} \cdot \nu^2 \tag{7.8}$$

$$\mathrm{RM} = \frac{e^3}{8\pi^2 \epsilon_0 m^2 c^3} \cdot \int_0^d n_e(s) B_{||}(s)\, \mathrm{d}s \tag{7.9}$$

Note that contrary to differential refraction discussed above, Faraday rotation depends on the absolute TEC. The rotation measure RM is $\sim$0.5 rad/m^2 for a typical nighttime value of the absolute integrated electron density of 10 TECU. Correction of Faraday rotation can be done by either directly measuring the rotation measure using known polarized sources/structures, or by combining Earth magnetic field models with independent TEC measurement, e.g. from GPS. This will be discussed in Sect. 7.3.2.

Differential Faraday rotation occurs when the product of TEC and magnetic field above two stations differ significantly, such that when the cross correlation of the two stations is calculated the signal is effectively rotated from the XX and YY correlations to XY and YX. This will be discussed in Sect. 7.3.1.

7.3.1 Differential Faraday Rotation

A secondary effect which is not only visible on polarized sources is differential Faraday rotation. This effect is due to the fact that different stations see different polarization rotation due to the spatial structure of the ionosphere, leading to a net rotation angle in the cross correlation. In terms of the measurement equation (recall Chap. 5, Eq. (5.1)), ignoring direction dependent effects:

$$\mathbf{V}_{ij}^{obs} = \mathscr{F}_{ij}\mathbf{V}_{ij}^{sky} \tag{7.10}$$

where $\mathscr{F}_{ij}$ is the Faraday rotation Mueller matrix composed of Jones matrices for stations i and j each with the form:

$$F_i = \text{Rot}\,\beta_i = \begin{pmatrix} \cos\beta_i & -\sin\beta_i \\ \sin\beta_i & \cos\beta_i \end{pmatrix}, \tag{7.11}$$

and where β_i is the average rotation angle as seen by station i. Differential Faraday rotation can be observed in the calibration solutions of all observations, but for low frequency data (LBA) and at times where the ionosphere is very wild it can be directly seen in the uv data. An example is shown in Fig. 7.4. For baselines longer than 10 km and frequencies below 80 MHz the effect cannot be ignored: the intrinsically unpolarized signal can even be fully rotated to the

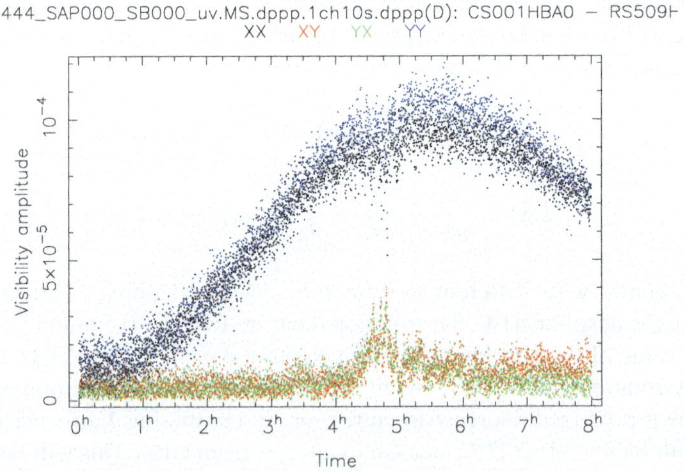

Fig. 7.4 Differential Faraday rotation visible in raw HBA uv data for the long core-remote baseline CS001HBA0 – RS509. The blue and black dots show the amplitudes of the XX and YY correlation versus time, the red and green are the XY and YX signals. For most of the observation the XY and YX signal is flat and noise-like, but an ionospheric event occurring just after $t = 4$ h causes significant leakage of the XX and YY signals into the cross correlations

XY and YX correlations, corresponding to a differential rotation angle of 0.5π. During calibration one should either solve for the full polarization matrix or include a rotation term as in Eq. (7.10). In BBS (Shulevski 2017) this can be accomplished by using the `CommonRotation` option in the model and to solve for the `CommonRotationAngle`. Note that this ignores the direction dependent effects of Faraday rotation.

Differential Faraday rotation provides a clean (i.e., only one parameter per station) independent measure of the variability of the ionosphere (+ Earth magnetic field). In principle, given a good knowledge of the Earth magnetic field along the line of sight, combined with the ΔTEC solutions derived from the calibration phases, it could be used to estimate the absolute TEC values by solving TEC in the following equation:

$$\Delta \mathrm{RM} = \Delta \mathrm{TEC} \cdot B_{||} + \mathrm{TEC} \cdot \Delta B_{||} \tag{7.12}$$

In practice, given the current accuracy of Earth magnetic field models, especially for the differential parallel magnetic field $\Delta B_{||}$, this turns out to be difficult. Only for the longest baselines (>80 km), where the $\Delta B_{||}$ values are significant (>1%) with respect to $B_{||}$, the method results in reasonable TEC values compared to GPS data.

7.3.2 RM Correction

For polarization measurements the main distortion comes from the time variability of the ionosphere. The TEC and magnetic field variation over the field of view can be ignored to first order, although, due to the $1/\nu^2$ dependency, the field of view variation of the rotation measure is significant at the lowest frequencies.

A first order correction to improve polarised signals can come from IONEX data (Schaer and Gurtner 1998), combined with Earth magnetic field models. The IONEX data, based on a fit of data of many satellites and GPS receivers, can be downloaded from several ftp servers.[4,5] The spatial resolution is of these data is typically 2.5°, and a new map is published every 1–2 h. It is also possible to download the GPS (RINEX) data itself, with much higher time resolution (∼30 s). The ALBUS software package[6] allows combination of the GPS data from several receivers to obtain TEC models with better time resolution. With this software it is also possible to estimate the RM as a function of time.

[4]ftp://ftp.aiub.unibe.ch/CODE/.

[5]ftp://gnss.oma.be/gnss/products/IONEX/.

[6]https://github.com/twillis449/ALBUS_ionosphere/.

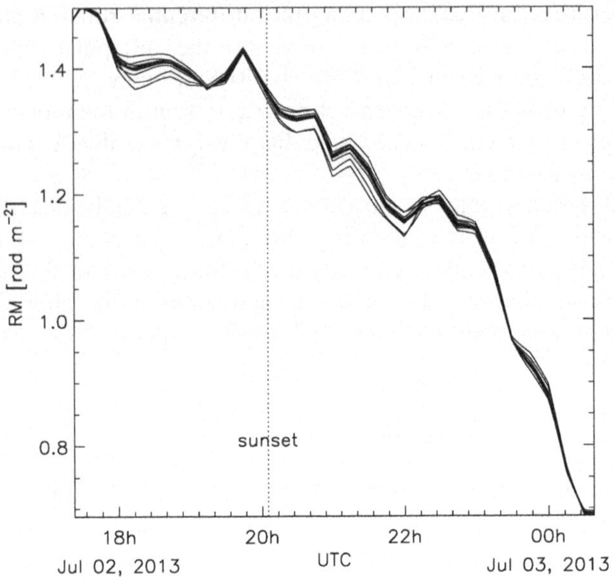

Fig. 7.5 GPS and WMM-deduced RM values for an HBA observation of the ELAIS-N1 field. Different lines correspond to different stations. Credit: Jelić et al., A&A, 568, 101, 2014, reproduced with permission © ESO

For the Earth magnetic field the two main global models are WMM[7] and IGRF.[8] To obtain the best estimate of the RM value, the correct approach would be to do some form of ray tracing through the ionosphere assuming a certain profile. This is the technique employed in the ALBUS software. The largest uncertainty comes from the IONEX data, which can have uncertainties up to 1 TECU, thus assuming a thin layer ionosphere at fixed height hardly increases the uncertainty on the RM values. The publicly available code RMextract[9] is designed to give a simple estimate of the TEC, vTEC, parallel magnetic field and RM value for a given observation. It also allows the user to generate an instrument table, which can be used to correct the data in BBS (Shulevski 2017), using the `FaradayRotation` option in the model.

The TEC and RM values deduced from CODE and WMM for an HBA observation of the Elais field taken from Jelić et al. (2014) are shown in Fig. 7.5. The need for a time dependent correction of the ionsopheric Faraday rotation is clear from this picture: the RM value changes by almost 1 rad/m^2 during the observation. In Fig. 7.6 the improvement in peak flux of a polarised source when applying these RM values to the data is shown.

[7]http://www.ngdc.noaa.gov/geomag/WMM/.

[8]http://www.ngdc.noaa.gov/IAGA/vmod/igrf.html.

[9]https://github.com/maaijke/RMextract.

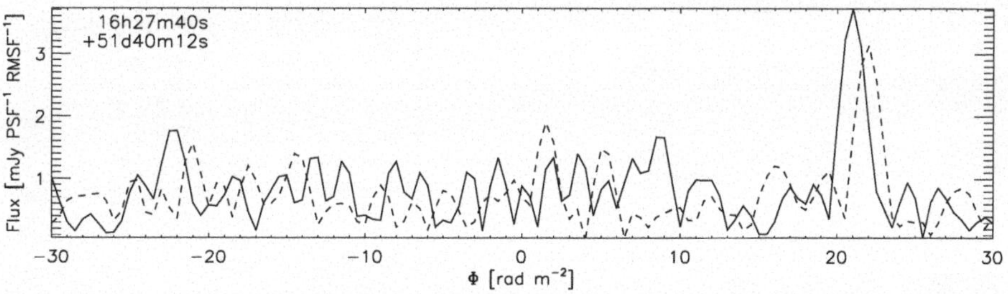

Fig. 7.6 Single line of sight through RM space at the position of a polarized source with intrinsic RM around 21 rad/m^2. Dashed line: uncorrected data, solid line: data corrected with the RM values from Fig. 7.5. The RM correction not only shifts the source in RM space but also focusses it, thereby increasing the peak flux by 20%. Credit: Jelić et al., A&A, 568, 101, 2014, reproduced with permission © ESO

RM calibration can be refined by making use the polarised signal in the data itself, using large bandwidth data. If the flux is high enough to allow high time resolution RM cubes of the polarised signal, the variation in time of these RM cubes must be mainly due to the ionosphere. The mean rotation measure of a polarised source (including the diffuse polarized Galactic foreground) can be fitted as function of time; see Chap. 10. Combined with the Earth magnetic field model this is another method to derive absolute TEC values from the data.

References

Heald, G.H., Pizzo, R.F., Orrú, E., et al.: Astron. Astrophys. **582**, A123 (2015)
Intema, H.T., van der Tol, S., Cotton, W.D., et al.: Astron. Astrophys. **501**, 1185 (2009)
Jelić, V., de Bruyn, A.G., Mevius, M., et al.: Astron. Astrophys. **568**, A101 (2014)
Koopmans, L.V.E.: Astrophys. J. **718**, 963 (2010)
Loi, S.T., Murphy, T., Cairns, I.H., et al.: Geophys. Res. Lett. **42**, 3707 (2015)
Schaer, S., Gurtner, W.: IGS Workshop Proceedings: 1998 Analysis Center Workshop (1998)
Shulevski, A. (ed.): The LOFAR Imaging Cookbook (2017)
Thompson, A.R., Moran, J.M., Swenson, G.W., Jr.: Interferometry and Synthesis in Radio Astronomy, 2nd edn. Wiley, New York (2001)

Chapter 8
Wide Field Imaging

Sebastiaan van der Tol

Abstract Visibilities cannot be fully corrected to compensate for direction dependent effects. Corrected visibilities are valid for one direction only. By contrast, images can be corrected in a direction dependent fashion. For example, an image can be adjusted to correct for the direction dependent sensitivity pattern of a primary antenna beam. However, this image-plane correction can only be properly applied if the beam is not time dependent. Effects that are both time and direction dependent can only be corrected by the imager, during the process of inversion from the visibility to the image domain. A-projection is one example of an algorithm that can apply these corrections efficiently. For LOFAR data, the AWimager software implements this algorithm. This chapter describes the A-projection algorithm, its implementation in AWimager, as well as tips for its usage and performance tuning.

8.1 Introduction

The LOFAR field of view is very large. While the exact size depends on observing mode and frequency (see van Haarlem et al. 2013), in common usage at HBA frequencies, the effective field of view is about 5° wide. Often one needs to image at least the first side lobe of the primary beam, as well, in order to deconvolve any bright sources that are located there. Imaging LOFAR data is usually by nature wide field imaging.

In the imaging process it is advantageous both mathematically and computationally to assume a flat (planar) sky. For wide fields of view, this assumption is not valid and leads to strong image artifacts. The W-projection (Cornwell et al. 2008) algorithm is an efficient method to take the curvature of the sky into account and has been adopted for use in many programs to perform interferometric imaging.

S. van der Tol (✉)
ASTRON, Dwingeloo, The Netherlands
e-mail: tol@astron.nl

G. Heald et al. (eds.), *Low Frequency Radio Astronomy and the LOFAR Observatory*, Astrophysics and Space Science Library 426,
https://doi.org/10.1007/978-3-319-23434-2_8

Due to the fact that LOFAR's aperture array stations are fixed to the ground and do not physically move to track the sky like dish-mounted receivers, the LOFAR beam changes substantially over time. Additionally, at low radio frequencies ionospheric effects can be severe, causing time and direction dependent phase errors (see Chap. 7). Because both of these effects are time variable, they cannot be fully compensated for after the imaging process by dividing out an averaged effective beam. The purpose of the A-projection algorithm (Bhatnagar et al. 2008) is to apply time and direction dependent corrections in the imager during the inversion process, where these effects can be properly treated.

The AWImager (Tasse et al. 2013) is the name of the imager that has been developed for the LOFAR pipeline and that implements both W-projection and A-projection. AWImager is built on top of the synthesis imaging code in CASA, the Common Astronomy Software Applications package.[1] It uses the CASA code for imaging and deconvolution steps, but implements specialized routines for gridding and degridding (see Sect. 8.4) that apply corrections for both the W-term and A-term. The A-term includes the LOFAR beam, as well as ionospheric effects.

Development of the AWImager software is still ongoing and the user interface is still subject to change. For up-to-date details on the usage of the AWImager, and the exact syntax of the various parameters, please consult the online documentation such as the LOFAR Imaging Cookbook.[2] This chapter focuses on the theoretical background needed to understand the meaning of the various parameters, as well as their effect on the resulting image quality and required processing time.

8.2 Data Model

Imaging can be seen as an estimation problem where the free parameters are the values of the pixels in the image. These values are estimated based on the observed data, the visibilities.

The equation describing the observed visibilities as a function of the unknown parameters is the Measurement Equation, which we write here as

$$V(u, v, w) = \iint_{S^2} I(\mathbf{s} - \mathbf{s}_0)A(\mathbf{s})e^{-2\pi i \nu \mathbf{b}\cdot\mathbf{s}/c}\,\mathrm{d}^2\mathbf{s} \tag{8.1}$$

where $\mathbf{s}$ is a vector indicating a direction on the sky that will be discussed further below, $\mathbf{b}$ is the baseline, ν is the observing frequency and c is the speed of light. The true intensity distribution on the sky is modeled as a continuous function $I(\mathbf{s})$ on a sphere. The reconstructed image will have the form of a rectangular, regularly

[1] http://casa.nrao.edu/.

[2] https://www.astron.nl/radio-observatory/lofar/lofar-imaging-cookbook.

sampled grid of points, representing the continuous function $I(\mathbf{s})$. The antenna response function (primary beam) is described by $A(\mathbf{s})$.

We will define a reference frame such that the baseline vector $\mathbf{b}$ in Eq. (8.1) can be written as $\nu\mathbf{b}/c = u\mathbf{u} + v\mathbf{v} + w\mathbf{w}$. Then the ideal relation between the reconstructed image and the true intensity distribution involves:

1. Projection onto a plane defined by two coordinates l and m, referred to as the "direction cosines".

$$I(l,m) = \frac{I(\mathbf{s})}{\sqrt{1-l^2-m^2}}$$

where

$$\mathbf{s} = l\mathbf{u} + m\mathbf{v} + n\mathbf{w}$$

and $n = \sqrt{1-l^2-m^2}$. The scaling is included in the definition of the projected image so that the integrated flux can be computed by integrating $I(l,m)$ over l and m.

2. Smoothing by convolution to a finite angular resolution,

$$I'(l,m) = I(l,m) * k(l,m)$$

where $k(l,m)$ is a smoothing kernel with an angular size given by the diffraction limited resolution of the synthesized aperture formed by the interferometer.

3. Discrete sampling of the image by simply recording the value of $I'(l,m)$ at a regular spacing of Δl and Δm (i.e., the pixel size),

$$I'[i,j] = I'(i\Delta l, j\Delta m)$$

where i and j are pixel indices.

The actual relation between the reconstructed image $I'[i,j]$ and the true intensity distribution $I(\mathbf{s})$ will only be an approximation of this ideal.

The discretized image can be written in a continuous form again by considering the pixel values to be described as a collection of point sources,

$$I'(\mathbf{s}) = \sum_{i=-X/2}^{X/2-1} \sum_{j=-Y/2}^{Y/2-1} I'[i,j]\delta\left(\mathbf{s} - \left(\mathbf{u}i\Delta l + \mathbf{v}j\Delta m + \mathbf{w}\sqrt{1-(i\Delta l)^2-(j\Delta m)^2}\right)\right)$$

where $\delta(x) = 0$ for non-zero values of x, and $\delta(0) = 1$. The image size is $X \times Y$ pixels.

Substituting this model into (8.1) yields the discrete version of the Measurement Equation,

$$V(u, v, w) = \sum_{i=-X/2}^{X/2-1} \sum_{j=-Y/2}^{Y/2-1} I'[i, j] e^{-2\pi \mathrm{i}\left(ui\Delta l + vj\Delta m + w\sqrt{1-(i\Delta l)^2-(j\Delta m)^2}\right)} \tag{8.2}$$

Note that in the description of the discrete model we have applied a smoothing filter that removed the high frequency components from the image. This puts a limit on the maximum baseline length for which the discrete Measurement Equation can be used.

8.3 Imaging as a Data Fitting Problem

There are different approaches to obtain the reconstructed image. Some are motivated by practical considerations while others have been developed on the basis of mathematical formalism.

A fairly straightforward approach is to recognize that under certain conditions the measurement equation resembles a two-dimensional Fourier transform. The Fourier transform is invertible, and there are computationally efficient techniques available to perform such an operation. With some adaptations a procedure similar to the inverse Fourier transform can be applied in order to construct an image from the visibilities. Additional tweaking of the procedure is usually used to improve the resulting image. For a comprehensive description of the details of the standard imaging technique, see Briggs et al. (1999).

An approach motivated from mathematical formalism would be to define a (statistical) metric of reconstructed image quality and derive the estimator that optimizes this metric. Such an approach usually leads to a problem that is unsolvable, either fundamentally or practically. Some ad hoc tweaking of the equations (metric, a priori information) is generally necessary to yield a practical solution.

In practice, both classes of technique lead to very similar methods. Most, if not all, involve the minimization of a weighted sum of the squared difference between the measurements and predicted visibilities.

The discrete form of the Measurement Equation can be compactly written as a matrix product

$$\mathbf{v} = \mathbf{A}\mathbf{i},$$

where $\mathbf{v}$ are the visibilities, $\mathbf{i}$ represents the source fluxes or pixel values, and $\mathbf{A}$ describes the relation between the two. In this framework, the imaging problem reduces to finding $\mathbf{i}$ given a (noisy) measurement of $\mathbf{v}$.

For a linear system of equations and in the presence of Gaussian noise, the optimal solution is the one that minimizes the weighted sum of the squared errors

$$\mathbf{i} = \arg\min_{\mathbf{i}} \left\| \mathbf{W}^{1/2} (\mathbf{v} - \mathbf{A}\mathbf{i}) \right\|^2$$

where $\mathbf{W}$ is a weighting function. The generic solution to this system of linear equations is given by

$$\mathbf{i} = \left(\mathbf{A}^{\mathbf{H}} \mathbf{W} \mathbf{A} \right)^{-1} \mathbf{A}^{\mathbf{H}} \mathbf{W} \mathbf{v}$$

where the superscript **H** indicates a Hermitian transpose operation.

Although this is a nice closed form solution to the theoretical imaging problem, in practice it is not very useful for two reasons.

The first problem is the size of the matrices. Computing $\mathbf{A}^{\mathbf{H}}\mathbf{A}$ requires $N_{pix} \times N_{pix} \times N_{vis}$ multiplications, where N_{pix} is the number of pixels ($X \times Y$) and N_{vis} is the number of visibilities. For a typical LOFAR HBA dataset the resulting image may cover a 4° field of view with 1″ pixels, requiring of order $10^4 \times 10^4$ pixels. Meanwhile, the number of visibility records corresponding to an 8-h observation with 1-s integrations and 62 stations is of order 10^8 per frequency channel. Thus for a single frequency, computing $\mathbf{A}^{\mathbf{H}}\mathbf{A}$ requires of order 10^{24} multiplications, which is impractical.

The second problem is that in virtually all cases the problem will be under-determined, which is to say that $\mathbf{A}^{\mathbf{H}}\mathbf{A}$ will not be invertible. One way to understand this is to recognize that there is an infinite set of images that all fit the data equally well. Extra constraints are needed to select the best image from this set.

In practice the minimization problem is solved by an iterative procedure where the current best guess for the image is updated based on the local gradient of a cost function. The update procedure either implicitly or explicitly forces additional constraints on the solution.

An example of such an iterative procedure is the CLEAN algorithm by Clark (1980):

1. Begin with an empty model image.
2. Construct a dirty (residual) image from the visibilities.
3. If a predefined threshold (e.g. the thermal noise level) or a maximum number of iterations has been reached, then stop.
4. Find peaks in the residual image and use these to define CLEAN components and update the model image. This step might use a secondary iteration called the minor cycle (see Cornwell et al. 1999).
5. Compute model visibility data from the CLEAN components accumulated so far.
6. Compute residual data by subtracting the model data from the observed data, and return to step 2.

At the end of this procedure, the CLEAN component model is convolved to match the angular resolution of the image and added to the final residual image.

It might not be immediately obvious that this algorithm is actually serving to minimize a cost function, by using the gradient to update the estimate. This is revealed by a close inspection of the gradient of the cost function, given by

$$\frac{\partial}{\partial \mathbf{i}} \|\mathbf{v} - \mathbf{A}\mathbf{i}\|^2 = \mathbf{A}^{\mathrm{H}}\mathbf{W}\left(\mathbf{v} - \mathbf{A}\mathbf{i}\right).$$

For an empty image ($\mathbf{i} = \mathbf{0}$) the righthand side reduces to $\mathbf{A}^{\mathrm{H}}\mathbf{W}\mathbf{v}$. That matrix product is the dirty image, as can be seen by considering the matrix formulation of the Measurement Equation. Meanwhile, the term $\mathbf{A}\mathbf{i}$ reflects the predicted visibilities for model image $\mathbf{i}$. As described above, the difference between the observed and predicted visibilities are the residual visibilities. So indeed the gradient of the cost function is the residual image. The implicit constraint that is applied through the update procedure is that the sky brightness distribution is composed of a finite number of point sources on an otherwise empty sky.

In the minor cycle, the point spread function (psf) is used to update the residual image. The psf can thus be seen as an approximation of the second order derivative of the cost function.

The iterative procedure avoids computing and inverting the matrix product $\mathbf{A}^{\mathrm{H}}\mathbf{A}$. For each iteration the residual visibilities and the residual image need to be computed.

A large class of image reconstruction algorithms rely on alternately evaluating the products $\mathbf{A}^{\mathrm{H}}\mathbf{v}$, i.e. making a dirty/residual image, and $\mathbf{A}\mathbf{i}$, predicting the visibilities for a given model image. The performance of these algorithms depends on the speed and accuracy of these basic operations. The next section describes the methods used to evaluate these products efficiently.

8.4 Gridding and Degridding

We will loosely follow the treatment of gridding and degridding provided by Thompson et al. (2001, Section 10.2). The extension to W-projection and A-projection follows naturally.

The discrete Measurement Equation (8.2) can be simplified somewhat by assuming $A(l, m) = 1$ (equal primary beam response in all directions) and that n is small,

$$V(u, v) = \sum_i \sum_j e^{-2\pi \mathrm{i}(ui\Delta l + vj\Delta m)} I[l, m]$$

For uv points on a regular grid this equation has the form of the Discrete Fourier Transform (DFT). The equation for the dirty/residual image then becomes the inverse DFT,

$$\hat{I}[l, m] = \sum_k e^{+2\pi i(ui\Delta l + vj\Delta m)} V(u_k, v_k) \tag{8.3}$$

if the measurements are on a regular grid. The DFT can be evaluated using the Fast Fourier Transform (FFT) which requires only $O(N_{pix} \log N_{pix})$ operations. This is a huge saving compared to the $O(N_{vis} N_{pix}^2)$ operations that are needed for a direct evaluation of (8.2) or its corresponding inverse operation to generate the dirty image.

The equation in Eq. (8.3) is directly applicable if the measured visibilities are on a regular grid of uv points. In this case, it would be straightforward to employ the computationally efficient FFT algorithm. In practice, the uvw coordinates of the visibilities are a continuous function of time and frequency (see, e.g., Thompson 1999). Therefore an intermediate step is required, taking the visibilities observed at continuous coordinates and putting them onto a regular grid of uv points. This step is called *gridding*. After this, an FFT operation can be used to obtain the image.

When computing model visibilities from a model image, for example during the CLEAN procedure, an FFT is used to obtain visibilities on a regular uv-grid. The model visibilities at the continuous uv coordinates matching the samples in the observed data are then computed from this grid. This step is called *degridding*. Figure 8.1 shows an example of a track of uv points.

The simplest way to put continuous visibility samples onto a regular uv grid would be to add each visibility to its nearest uv grid point. Of course this is not an exact procedure, and the rounding that is required will have an effect on the final image that is difficult to deal with. That is because rounding is a non-linear operation. Perhaps surprisingly the operation of resampling to the nearest grid point can be expressed as a sequence of linear operations. This simplifies the analysis of the effect of gridding on the image.

Let us first recall one of the fundamental properties of the Fourier transform. A property that is a core concept in radio interferometry is the convolution theorem, which states that the convolution of two functions in one domain is equivalent to the multiplication of the Fourier transforms of the functions in the inverse domain, and vice versa. Many operations can be written as either a convolution or a multiplication by some particular function. Some key examples of Fourier transform pairs, that are relevant for gridding and degridding, are summarized in Fig. 8.2.

A visibility sample can be modeled as a Dirac delta function, or "pulse," at its uv-location multiplied by its (complex) value. Gridding this visibility to the nearest grid point can be represented as a convolution with a 2D box function, followed by a multiplication with the comb function (see the third row of Fig. 8.2). The convolution with the box function changes the Dirac pulse to a flat-topped box that overlaps exactly one grid point, namely the one that is closest to the visibility sample. The subsequent multiplication by the comb function (pulses at the locations of the grid points) reduces the box to a Dirac delta function again, but now situated at the location of the nearest grid point. See Fig. 8.3 for an illustration of this sequence of operations.

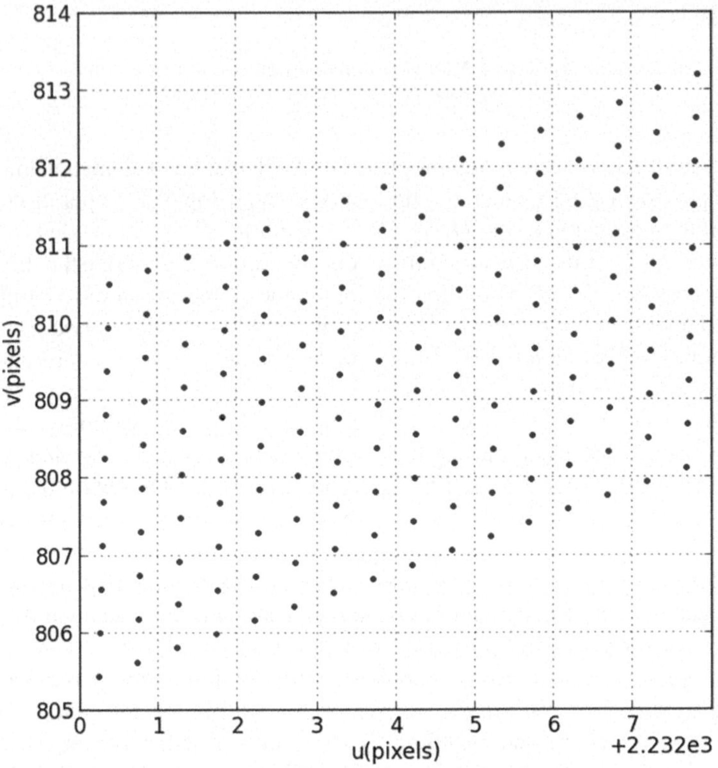

Fig. 8.1 An example uv track. The dots represent the location in the uv plane of visibility measurements for a single baseline. The uv coordinate is measured in wavelengths, hence the samples for different frequency channels at the same time slot lie on a line through the origin. In this way we see that for this set of visibility samples, time varies along the vertical "track" and for each time sample, frequency measurements are found along diagonal lines to the upper right. Note that each visibility is an integral over a small time and frequency interval, so each measured visibility is an average over a region in the uv plane

Now let us examine the effect of this procedure in the image domain. The box function employed in the gridding procedure is illustrated in one dimension in the bottom left of Fig. 8.2. Following the convolution theorem, the convolution of the visibility with a box function in the uv domain is equivalent in the image domain to a multiplication with the Fourier transform of the box function. The Fourier transform of the box function is the sinc function and is illustrated in one dimension in the bottom right of Fig. 8.2.

Likewise, the subsequent multiplication with the sampling function in the uv domain corresponds to a convolution in the image domain with the Fourier transform of the sampling function. The Fourier transform of the sampling function, or Dirac comb function, is another comb function but with a spacing given by the inverse of the spacing in the conjugate domain. Since the spacing of uv grid points

<table>
<tr><th>$F(t)\cdot G(t)$</th><th>$F'(f) * G'(f)$</th></tr>
<tr><td>Dirac Delta</td><td>1</td></tr>
<tr><td>Shah / Dirac Comb / bed of nails
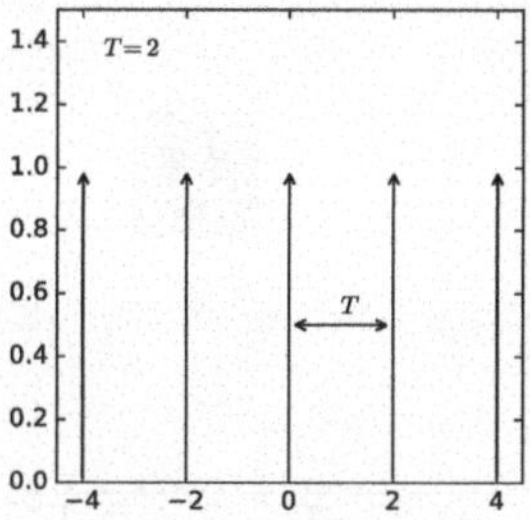

$\text{III}(t/T) = \sum_{k=-\infty}^{\infty} \delta(t/T - k)$</td><td>Shah / Dirac Comb / bed of nails
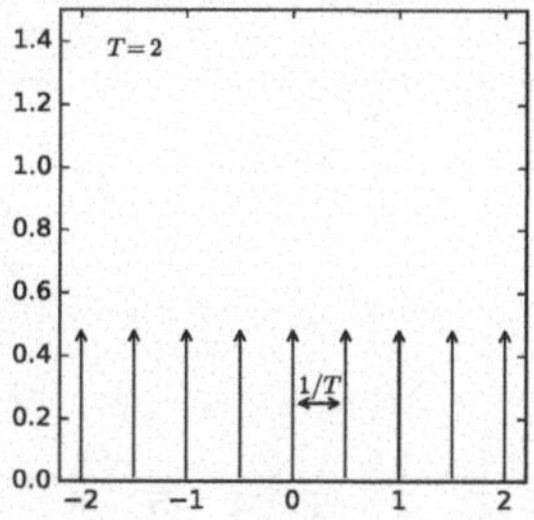

$\frac{1}{T}\text{III}(fT) = \frac{1}{T}\sum_{k=-\infty}^{\infty} \delta(fT - k)$</td></tr>
<tr><td>Box
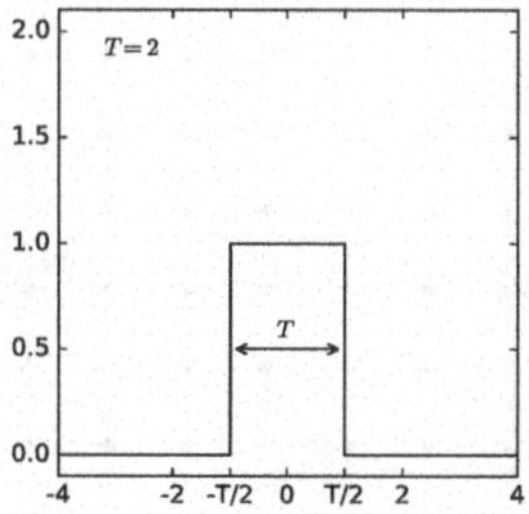

$\text{rect}_T(t) = \begin{cases} 1 & \lvert t\rvert \le T/2 \\ 0 & \lvert t\rvert > T/2 \end{cases}$</td><td>Sinc
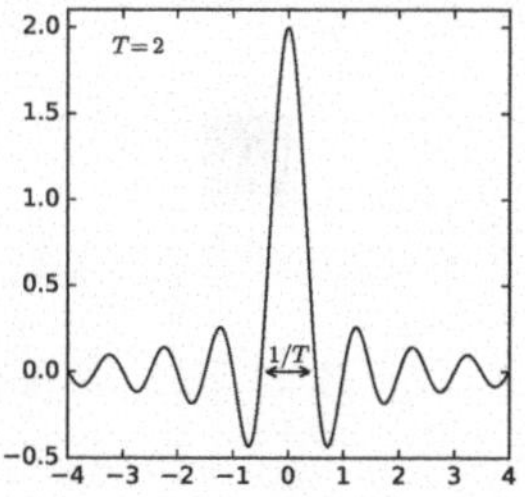

$T\,\text{sinc}(fT) = \frac{T\sin(\pi f T)}{\pi f T}$</td></tr>
</table>

Fig. 8.2 Fourier transform pairs. The top row recalls the convolution property of the Fourier transform. The remaining rows express the representation of a key function in the time domain on the left, and the representation of the Fourier transform pair in the conjugate (frequency) domain on the right

is inversely proportional to the field of view of the image, the comb function that is applied in the image plane has a spacing proportional to the field of view. In effect this sequence of operations folds sources located outside the image into the image, as is shown in Fig. 8.4.

Aliasing of sources outside the field of view into the image is obviously not desirable, so this discussion immediately shows that the box function is not a good choice for the procedure illustrated in Fig. 8.3. The main issue is the high sidelobes of the Fourier transform of the box function, which permit sources outside of the field of view to be folded into the image. On the other hand, the advantage of formulating the gridding procedure as a convolution now becomes clear. All that we have to do is replace the box function with a function that has a Fourier transform pair with more desirable properties.

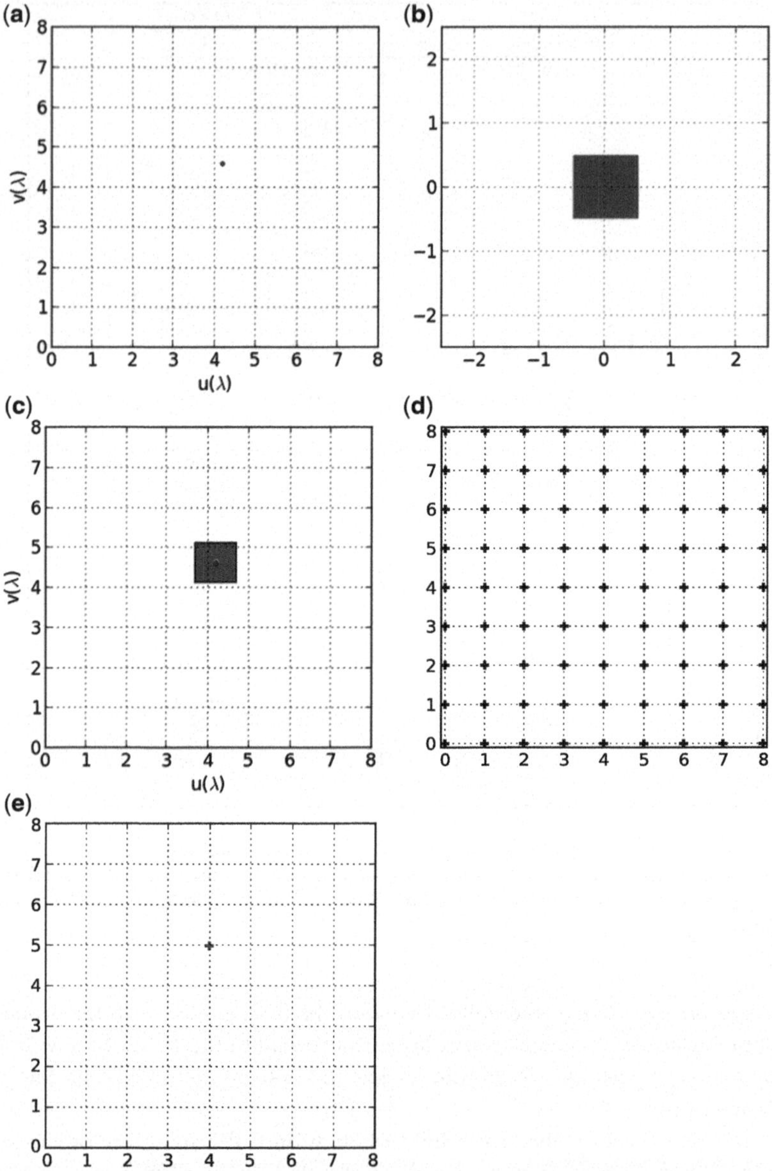

Fig. 8.3 Gridding of a single sample to the nearest pixel in steps. A single sample is modeled as a Dirac delta function at location (u, v). The sample is convolved by a 2D box function with size equal to that of the pixels. The resulting box overlaps with exactly one grid point. After multiplication by a grid of Dirac delta pulses only the pulse at the overlapping grid point remains. Effectively the sample has been moved to the nearest grid point. (**a**) A Dirac delta function at the position of the visibility in the uv plane. (**b**) The 2D box function. (**c**) The convolution of the Dirac delta function (8.3a) and the 2D box (8.3b). (**d**) The 2D comb function. (**e**) The product of the result of the convolution (8.3c) and the 2D comb function (8.3d). The visibility in 8.3a has been "gridded"

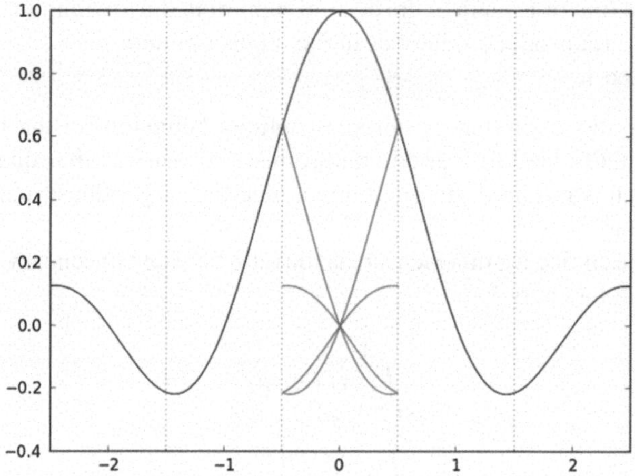

Fig. 8.4 Box function and aliasing. The Fourier transform of a box function is the sinc function (blue curve). Convolution in the uv domain by a box function is equivalent to multiplication in the image domain by a taper with considerable sidelobes. Subsequent sampling in the uv domain causes the sidelobes to show up in the main image (green curves)

The convolution function needs to have a Fourier transform with very low sidelobes (referred to as providing little "support" outside of the image). The function also needs to have limited support in the uv domain in order to limit the number of affected pixels in the uv grid, and thus to limit the required number of operations. The box function that we have already used clearly satisfies the second criterion since by design it affects only a single pixel in the uv domain. However as we have seen, its behaviour in the image domain is poor. By contrast, the ideal function in the image domain would be a box function with unit response within the field of view, and zero response outside. However in this case the convolution function would then be the sinc function, which would provide infinite support in the uv domain and effectively smear each visibility throughout the uv plane.

Clearly, no function can fully satisfy the requirements for ideal behaviour in both the image domain and the uv domain. Limited support in the uv domain means infinite support in the image domain and vice versa. Thus, there will always be some aliasing in the image formed through this technique, but the gridding procedure can be tuned to provide much better properties than the box function.

An ad hoc procedure to generate a better convolution function is to start with a box function of the desired width in the uv domain. It will have sidelobes in the image domain, which can be removed by following an iterative procedure.

1. Transform the convolution function to the image domain.
2. Truncate the function in the image domain to suppress support outside of the field of view. The resulting function will now have sidelobes in the uv domain.
3. Transform the function back to the uv domain.

4. Truncate the function again to the desired support in the uv domain. The resulting function will reintroduce sidelobes in the image domain.
5. Return to step 1.

The procedure above alternately enforces optimal behaviour in the two domains. Will this eventually lead to a good compromise? To answer that question a more formal approach is required, first defining a measure of goodness and then finding the best function.

A reasonable choice for this measure is the ratio between the energy in the desired region and the total energy:

$$\frac{\int_{-A}^{A} |f(t)|^2 dt}{\int_{-\infty}^{\infty} |f(t)|^2 dt} \tag{8.4}$$

By this criterion the optimal convolution function is the one that minimizes the energy in the sidelobes in the image domain, under the constraint that its support in the uv domain does not exceed the chosen limit. This problem has been studied in detail by Slepian and Pollak (1961) and the solution to the minimization problem is the Prolate Spheroidal Wave Function (PSWF).

Interestingly the PSWF is the eigenfunction with the largest eigenvalue of the ad hoc procedure described above, as determined by a power iteration, i.e. repeatedly applying the operation to a random starting vector. Thus the ad hoc procedure of repeatedly transforming from one domain to the other and truncating to the desired support will in fact eventually converge to the optimal convolution function.

CASA makes use of a PSWF that is seven pixels wide in the uv domain. The spheroidal function is zero outside the region of interest $[-0.5, 0.5]$.

Note that the application of the spheroidal convolution function applies a multiplicative taper towards the edges of the image, as can be see in Fig. 8.5. To obtain the final image, the taper function needs to be divided out again. However, the near-zero value of the function near the edges causes an amplification of the numerical noise. Therefore some padding of the image size is required, so that the outer region can be discarded.

8.4.1 Sampling of the Convolution Function

In practice, the convolution function is not evaluated at the required continuous coordinates during the gridding process. Instead, to gain computational efficiency the convolution function is precomputed on a grid, and during gridding the nearest sample is looked up. This process is illustrated in Fig. 8.6. The grid that is sampled for the precomputed convolution function is denser than the uv grid that is used

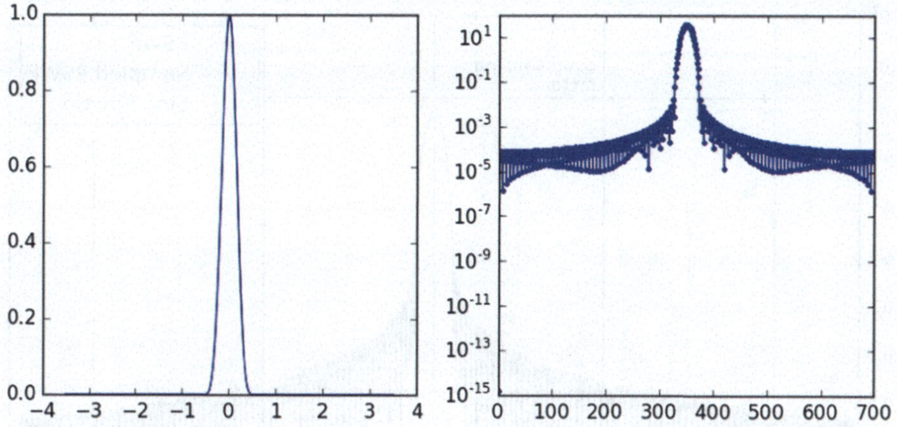

Fig. 8.5 The prolate spheroidal wave function. Left: the spheroidal function. Right: the Fourier transform of the spheroidal function

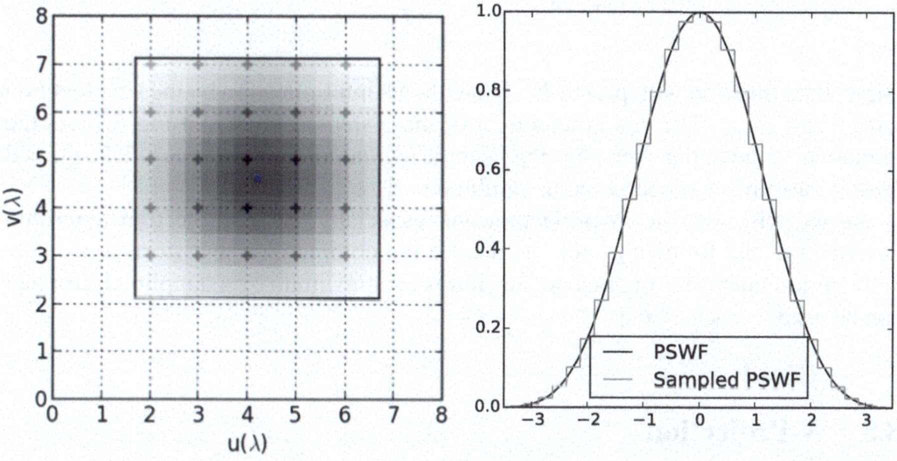

Fig. 8.6 The convolution function is precomputed on a grid. This grid has a higher resolution than the uv grid by the oversampling factor

to construct the image by FFT. The ratio between the sampling of the convolution function and that of the uv grid is called the oversampling factor.

How is the oversampling factor (N_{over}) determined? Effectively the convolution function is replaced by a stepwise approximation, as shown in Fig. 8.6. The discontinuities of this stepwise approximation cause sidelobes in the Fourier transform, or equivalently aliasing in the image domain. The stepwise approximation of the convolution function can be seen as a multiplication of the spheroidal function by the Dirac comb function with period $1/N_{\text{over}}$ and then a convolution by a box function with width $1/N_{\text{over}}$. In the image domain that is a convolution by the

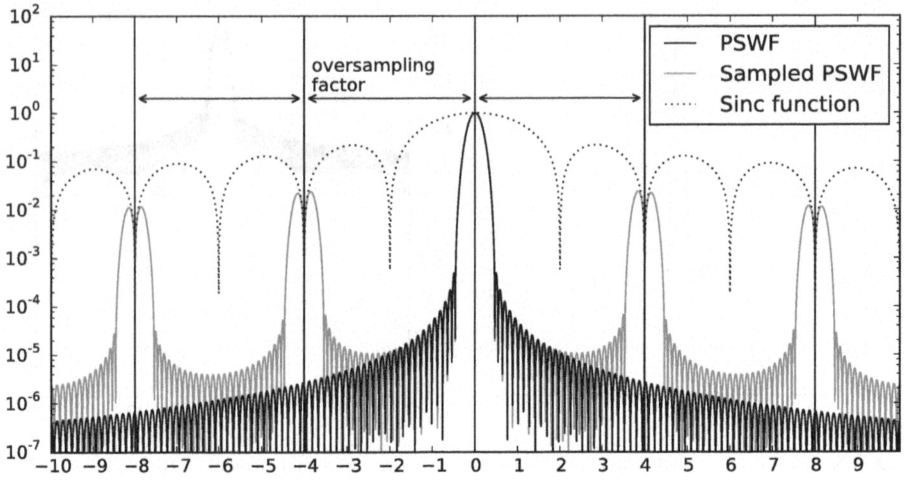

Fig. 8.7 Illustration of the aliasing induced by the gridding procedure, and its suppression by oversampling the convolution function

Dirac delta function with period N_{over} and then a multiplication by the sinc function with width N_{over}. The combined effect of these steps is shown in Fig. 8.7. As the numerical value of the oversampling factor is increased, the sidelobes will be pushed farther outwards and decreased in amplitude.

As we will see, the W-projection and especially the A-projection algorithm described in the following sections require the computation and storage of many convolution functions. In practice, this limits the maximum oversampling factor that can be used.

8.5 W-Projection

We have so far restricted ourselves to the use of the FFT only in cases where the product nw is small. In the previous sections it was shown that gridding and degridding inevitably leads to a multiplication in the image domain. We need to make sure the function behaves well in both the image and the uv domain, but other than that we are free to choose a suitable convolution function. The convolution functions that we have explored so far are simple tapers designed to have a minimal impact on the image. We could however choose functions to apply a non-negligible, but desired, effect in the image domain.

Let us examine the Measurement Equation again. As we have noted earlier, it is nearly a Fourier transform. Rearranging the Measurement Equation we have

$$V(u, v, w) = \int\int I(l, m)e^{-2\pi i n w}e^{-2\pi i(ul+vm)}dl dm$$

and from this formulation we can see that it is the Fourier transform of the image, multiplied by an extra term

$$V(u, v, w) = \mathscr{F}(I(l, m))\, e^{-2\pi i n w} \tag{8.5}$$

If we choose to incorporate this multiplicative term, the so-called W-term, as part of the convolution function, then imaging using the FFT can also be used in case nw is not negligible. Examples of this condition are for large fields of view, and/or inclusion of long baselines. A full description of the W-projection algorithm is given by Cornwell et al. (2008).

Because the W-term includes

$$n = 1 - \sqrt{l^2 + m^2},$$

support of the convolution function grows with the square of the angular diameter of the image. This can be problematic and we will address this in the next section.

8.6 W-Stack Algorithm

The support of the convolution function due to the W-term can grow very large. For example it would not be unusual in LOFAR imaging applications for the convolution function to reach hundreds of pixels in one dimension. This leads to a considerable computational cost. There are several ways to reduce the support of the convolution function.

1. **Facets.** An image can be broken into smaller sub-images, or facets. For each of the individual facets, the support of the convolution function is substantially reduced.
2. **Snapshots.** Instead of imaging the full uv tracks all together, sub-images can be formed from subsets of the visibilities over short periods in observing time called snapshots. Within each snapshot, the rotation of the array will be small. Since all antennas in an array typically lie approximately on a plane, the w coordinate will be small for images projected onto the tangent plane at zenith. After reprojection to the desired tangent plane, each snapshot can be added to a final cumulative image.
3. **W Stacking.** In the W Stacking approach described by Cornwell and Humphreys (2011, Section 8.3), the data are not gridded onto the $w = 0$ plane as is implicitly done by writing Eq. (8.5), but instead to a plane closest to each visibility in w distance. This is illustrated schematically in Fig. 8.8.

The AWImager used for LOFAR imaging uses W Stacking, as does another modern and widely used imaging application called `wsclean` (Offringa et al. 2014). For each w plane, a Fourier transform and corresponding image-plane correction needs to be computed. Despite these additional costs, the visibilities can

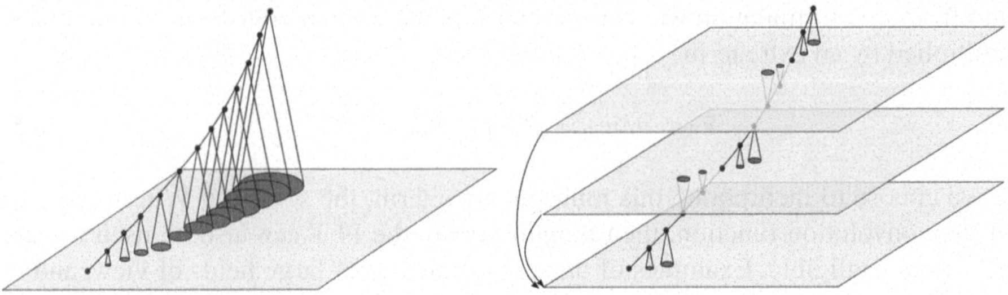

Fig. 8.8 W-Stacking algorithm. For large w coordinates the convolution function becomes excessively large. The W stacking algorithm reduces the size of the convolution function by employing multiple grids each with a different w offset. Visibilities are gridded to the grid with w offset nearest to the w coordinate of the visibility. After gridding, each grid is then convolved with the convolution function corresponding to the w offset and then added to the $w = 0$ grid. When a sufficiently large number of visibilities are gridded per grid, this is more efficient than using the larger convolution function per visibility

be gridded with smaller convolution functions which typically leads to improved performance.

For practical reasons, the different w planes cannot be simultaneously kept in memory. Therefore the AWImager steps through chunks of visibilities and within each chunk steps sequentially through the w planes, gridding the visibilities that are in memory at that moment and that belong to the current w plane. The Fourier transform and image-plane correction are performed, and the w plane is added to the cumulative image. After completing the iteration over all w planes, the next chunk of visibilities is read and the loop over the w planes restarts. This sequential behaviour means that for each chunk of visibility data, the Fourier transforms and image-plane corrections are repeated. It is therefore advantageous to choose the visibility chunk size as large as allowed by the available memory.

8.7 A-Projection

In Sect. 8.5 we have seen that a term in the Measurement Equation that is not part of the Fourier transform can be factored out and included in the convolution function (see Eq. (8.5)). This clever use of the convolution function is not limited to the W-term. We can also include the effect of the primary beam ("A-term") in the convolution function (CF):

$$CF(u, v) = \mathscr{F}\left(A_1(l, m)A_2^*(l, m)W(l, m)\mathrm{Sph}(l, m)\right) \tag{8.6}$$

where $A_1(l, m)$ and $A_2(l, m)$ are the primary beam functions appropriate for antennas 1 and 2, $W(l, m)$ is the contribution to the convolution function for a given w plane from the W-term, and $\mathrm{Sph}(l, m)$ is the PSWF discussed in Sect. 8.4.

Now that the convolution function not only depends on the W-coordinate, but also on the (time varying) beam of the two stations in the baseline, there are many more convolution functions to be computed. The time interval for which the convolution functions are recomputed is controlled in the AWImager by the `gridding.timewindow` parameter. Depending on the parameters of the computation of the convolution functions, those operations can dominate over the actual (de)gridding.

During gridding with A-projection, the visibility data are multiplied by the conjugate of the beam. Multiplying by the inverse of the beam is undesirable in the case of a time-variable beam, because it effectively puts the most weight on the visibilities that contain the least information. The conjugate negates the phase of the beam, canceling it in the multiplication, so that all contributions are summed coherently. The beam amplitude is applied twice, once by the instrument itself and once by the convolution function. The noise on the other hand is multiplied by the beam only once, by the convolution function.

Effectively, both the source brightness distribution and the noise *power* are multiplied by the mean squared beam. Normalizing the image by the root mean squared (rms) beam yields a flat noise image—in other words, the noise power is constant over the image, but the source power is suppressed by the rms beam. Normalizing the resulting flat noise image once more by the rms beam yields a flat gain image—in other words, the gain is constant over the image and the source brightnesses are corrected, but the noise is enhanced by the rms beam.

Most deconvolution algorithms expect a flat noise residual image, so that is what AWImager provides to the deconvolution algorithm. To compute the residual visibilities by subtracting the contents of the resulting model image, it is first normalized to a flat gain image. After deconvolution, the final (restored) image is normalized to a flat gain image. The mean squared beam that is used to normalize the image is computed on the fly by summing all the squared A-terms.

8.8 Estimating the Effect of the Gridding Parameters on the Final Image

In setting the parameters used by AWImager for (de)gridding, a trade-off has to be made between image quality on the one hand, and runtime and/or memory usage on the other hand. Since the different gridding parameters and the observational parameters interact in complicated ways it is not always possible to predict their effect on the final image quality. Meanwhile, testing a large range of settings is too time consuming. The artifacts introduced by the approximations in the gridding process only become apparent after deconvolution. This section describes methods to assess the impact of the gridding parameters on the final image quality, without the need to perform deconvolution. Of course this does not address other sources of image errors such as calibration errors, deconvolution artifacts, beam model errors

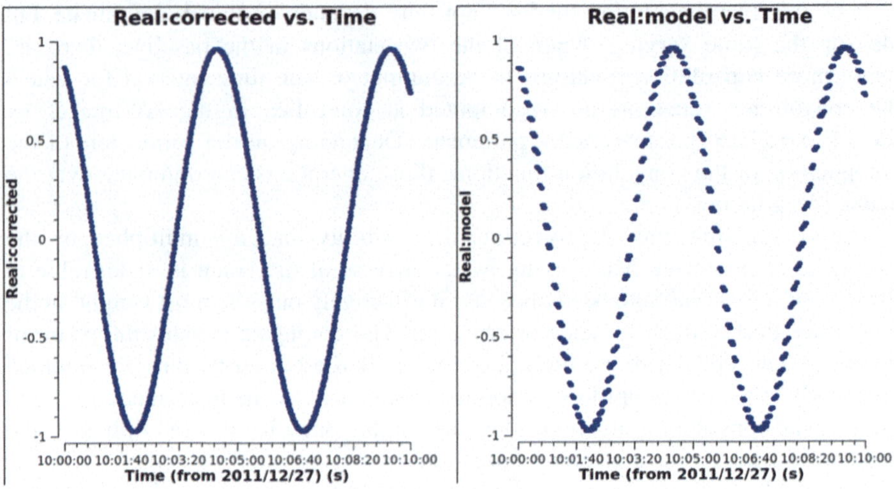

Fig. 8.9 Comparison between visibilities predicted by DPPP (left) and by AWImager (right). The differences are caused by the approximations made in the (de)gridding process

and thermal noise. The gridding parameters are chosen to optimize computational performance while not contributing significantly to the final image noise.

Previously in this chapter, imaging was framed as a minimization problem. In the case of noise-free data, perfect calibration, and a sky where all sources are pointlike and coincide exactly with the grid points of the image, the sky can be represented perfectly by the image and the cost function and its gradient, i.e. the residual image, will be zero. Any deviation from zero can then solely be attributed to approximations made during the gridding and degridding process. To first order, the residual between a model image and an image formed from the corresponding simulated data for this model, represents the error that gridding adds to the final image.

To assess the impact of the gridding parameters on the final image, we require a procedure that is quicker than simply running a complete deconvolution and inspecting the result. Waiting for CLEAN to converge is too time consuming to efficiently study the effect of the parameters on the quality of the final map. Meanwhile, subtracting the model data is the most critical step of clean and the accuracy of this step can be seen immediately on small subsets of the data. Therefore an effective and efficient approach is to compare degridded (predicted) visibilities to an exact evaluation of the Measurement Equation, for example by DPPP. An example of such a comparison is provided in Fig. 8.9.

Here, we provide a step-by-step procedure to follow this approach on any visibility dataset. The concept is to create a residual image from simulated data with the exact sky model used to generate the data as model image. Ideally the resulting image should be exactly zero. The noise level in this image indicates how much noise is added by inaccuracies in the (de)gridding step.

We start with an empty measurement set. This could be (a copy of) an existing measurement set, or created from scratch using for example the `makems` tool, using a configuration file like the one reproduced below (called `makems.cfg`):

```
StartFreq=120e6
StepFreq= 0.5e6
StartTime=2011/12/27/10:00:00.000
StepTime=1
RightAscension=23:23:22.7600
Declination=+58.50.16.3000
NBands=1
NFrequencies=1
NTimes=600
NParts=1
TileSizeFreq=8
TileSizeRest=10
WriteAutoCorr=T
AntennaTableName=ANTENNA
MSName=testdata.MS
```

Next an empty image is needed which can be created by

```
awimager2 operation=empty data.ms=testdata.MS image.npix=1000
    image.cellsize=10arcsec output.imagename=model.img
```

To create an extremely simple model, a single point source is added by setting a single pixel to a non-zero value using the `ia` tool in `casapy`:

```
ia.open("model.img")
pos = [0,0,1000,1000]
r = rg.box( blc = pos, trc = pos )
ia.set(pixels=1.0, region=r)
```

A source catalog file in text format can be created from this image with

```
casapy2bbs.py model.img model.catalog
```

From the resulting text file, a sourcedb for use in BBS can be created with

```
makesourcedb in=model.catalog out=model.sourcedb format='<'
```

The sourcedb can be used as input for DPPP to compute simulated visibilities, using the following parset (`dppp.parset`):

```
msin=testdata.MS
msout=.
msout.datacolumn=CORRECTED_DATA

steps=[predict, applybeam]

predict.type=predict
predict.sourcedb=model.sourcedb
predict.usebeammodel=true    # Apply a beam model for each
    source
applybeam.type=applybeam   # Correct for the beam in the phase
    center
```

Predict the data from the model image using the AWImager

```
awimager2 operation=predict data.ms=testdata.MS model=model.img
     gridding.padding=1.5 gridding.oversample=3
```

Plot the residual data

```
plotms(vis="testdata.MS",yaxis="real",ydatacolumn="residual",
    antenna="0&50",correlation="XX",
    plotfile="oversample3padding1.5.png", overwrite=True,
    plotrange=[0,0,-1,1])
```

Create the residual image

```
awimager2 operation=image data.ms=testdata.MS
    image.type=residual image.npix=4000 image.cellsize=5arcsec
    output.imagename=residual_oversample15_padding1.5
    gridding.padding=1.5 gridding.oversample=15
```

Compute the rms value of the residual image using capapy imstat

```
print "%.1e" % imstat(imagename = '
    residual_oversample15_padding1.5.flatnoise')['rms'][0]
```

The resulting rms residual noise levels that remain after running this test for various values of the oversample factor are shown in Table 8.1. As the oversample factor increases, the image artifacts are reduced but the AWImager will require more resources and take longer to run.

Table 8.1 Resulting rms values from residual images produced using different values of the AWImager oversample factor

Oversample factor	rms residual
3	3.5e−04
7	1.1e−04
15	6.3e−05

References

Bhatnagar, S., Cornwell, T.J., Golap, K., Uson, J.M.: Astron. Astrophys. **487**, 419 (2008)

Briggs, D.S., Schwab, F.R., Sramek, R.A.: In: Taylor, G.B., Carilli, C.L., Perley, R.A. (eds.) Synthesis Imaging in Radio Astronomy II. Astronomical Society of the Pacific Conference Series, vol. 180, p. 127 (1999)

Clark, B.G.: Astron. Astrophys. **89**, 377 (1980)

Cornwell, T.J., Humphreys, B.: Analysis of convolutional resampling algorithm performance, SKA memo 132, Technical report, SKA Organisation (2011)

Cornwell, T., Braun, R., Briggs, D.S.: In: Taylor, G.B., Carilli, C.L., Perley, R.A. (eds.) Synthesis Imaging in Radio Astronomy II. Astronomical Society of the Pacific Conference Series, vol. 180, p. 151 (1999)

Cornwell, T.J., Golap, K., Bhatnagar, S.: IEEE J. Sel. Top. Signal Process. **2**, 647 (2008)

Offringa, A.R., McKinley, B., Hurley-Walker, N., et al.: Mon. Not. R. Astron. Soc. **444**, 606 (2014)

Slepian, D., Pollak, H.O.: Bell Syst. Tech. J. **40**, 43 (1961)

Tasse, C., van der Tol, S., van Zwieten, J., van Diepen, G., Bhatnagar, S.: Astron. Astrophys. **553**, A105 (2013)

Thompson, A.R.: In: Taylor, G.B., Carilli, C.L., Perley, R.A. (eds.) Synthesis Imaging in Radio Astronomy II. Astronomical Society of the Pacific Conference Series, vol. 180, p. 11 (1999)

Thompson, A., Moran, J., Swenson, G.: Interferometry and Synthesis in Radio Astronomy. Wiley, Hoboken (2001)

van Haarlem, M.P., Wise, M.W., Gunst, A.W., et al.: Astron. Astrophys. **556**, A2 (2013)

Chapter 9
The LOFAR Standard Imaging Pipeline

George Heald

Abstract The Standard Imaging Pipeline (SIP) is LOFAR's automatic system for the calibration and imaging of interferometric data. It is designed to be generally applicable to all imaging observations and to provide a reasonable quality of output data product in most cases. The SIP is composed of combinations of the individual calibration and imaging programs covered elsewhere in this lecture series. In this lecture the overall strategy of the SIP will be presented, along with information about how to set up calibration and imaging jobs, a description of ongoing work to improve the pipeline, and illustration through example datasets.

9.1 Context

The LOFAR telescope is capable of producing a tremendous volume of data. Normal imaging observations routinely produce several tens of terabytes of raw visibility data. Specialized techniques are required to produce high-quality image products that can be used to produce science publications. The prospect of turning several terabytes of uncalibrated visibility data into a single high-quality image can be a rather daunting task, particularly for the vast majority of LOFAR users that do not have long and detailed experience with the requisite low-level software routines. The LOFAR Standard Imaging Pipeline (SIP) is a very high-level software tool intended to allow users to automatically produce calibrated visibilities and image products from normal interferometric observations. As with any automated system, the outcome cannot be expected to be perfect, but it should provide a reasonable starting point in most cases. The aim of this chapter is to explain in detail what the

G. Heald (✉)
ASTRON, Dwingeloo, The Netherlands

CSIRO Astronomy and Space Science, Kensington, WA, Australia
e-mail: george.heald@csiro.au

G. Heald et al. (eds.), *Low Frequency Radio Astronomy and the LOFAR Observatory*, Astrophysics and Space Science Library 426,
https://doi.org/10.1007/978-3-319-23434-2_9

SIP does, what the assumptions are, and under what circumstances good results can be expected.

9.2 Goals and Assumptions

The goal of the SIP is to facilitate the production of calibrated visibilities and, from these, images that are of sufficient quality to enable scientific analysis. This "science readiness" constraint is rather subjective and depends on a wide range of factors, not least being the definition of the science question at hand in a given project. Nevertheless, image properties corresponding to the statistics listed in Table 9.1 are typically used as the overarching goal of SIP processing. This set of goals is roughly based on the ambitions of the Survey Key Science Project (KSP)'s shallow Tier (LOFAR Two-metre Sky Survey or LoTSS; Shimwell et al. 2017), and images with those properties are (not coincidentally) adequate to enable detailed study of individual objects, while providing resolved imagery of a vast number of radio galaxies in the typical field of view.

The design and realization of a pipeline that is capable of producing images of this quality is by no means a trivial endeavour, and it is made harder if the stated goals must be achieved under all possible circumstances. Therefore it is important to set boundary conditions within which the goals are feasible. For LOFAR, these conditions include:

1. Lack of extremely bright sources in the field of view
2. Reasonably high elevation

Table 9.1 SIP goals

Property	Value (HBA)	Value (LBA)
Integration time	8 h	8 h
Bandwidth	48 MHz	48 MHz
Image depth[a]	100 μJy beam^{-1}	1 mJy beam^{-1}
	$\sim 4 \times \sigma_{\text{thermal}}$	$\sim 2 \times \sigma_{\text{thermal}}$
Resolution[b]	10″	30″
Field of view[c]	11 □° (HBA_DUAL_INNER)	12 □° (LBA_OUTER)
	≅ 9000 × 9000 3″ pixels	≅ 3000 × 3000 10″ pixels
Dynamic range	$5 \times 10^4 : 1$	$1 \times 10^4 : 1$
	for a 5 Jy source	for a 10 Jy source

[a]Estimated thermal noise (σ_{thermal}) values calculated using this sensitivity calculator: https://support.astron.nl/ImageNoiseCalculator/sens.php and assuming use of the HBA_DUAL_INNER and LBA_OUTER modes respectively, with all Dutch stations
[b]Target resolution; does not reflect the highest resolution accessible with the data
[c]Sky area within the nominal HPBW of the station beam in HBA_DUAL_INNER and LBA_OUTER modes, from van Haarlem et al. (2013), at reference frequencies of 150 and 60 MHz respectively. Also provided are the number of pixels needed to image at the target resolution to a diameter twice the station beam HPBW

3. Normal ionospheric activity
4. Standard primary calibration setup

The first of these conditions is related to the desired dynamic range listed in Table 9.1. If the target field contains one or more sources substantially brighter than can be found in the average region of sky, good quality images may be difficult to achieve. In particular, very bright and complicated objects like the "A-team" sources are highly problematic; sources in the next fainter tier (typically extended 3C sources) can also cause severe issues.

The second condition reflects the fact that for an aperture array like LOFAR, the sensitivity pattern fixed on the sky by the element beam results in a loss in sensitivity at low elevation. Moreover, the uncertainties in the beam model are larger toward the horizon. Finally, observations at low elevation cause the projected station size to shrink, which has the effect that the station beam grows, requiring a larger field of view to be imaged. Observations in that part of the sky may therefore be limited by systematic effects. Observations down to the celestial equator have proven to be well behaved, and it is expected that imaging as far south as $\delta \approx -20°$ is possible, but the SIP parameters may need to be adjusted.

The third condition reflects the fact that the SIP calibration procedure cannot function properly if the ionospheric conditions are bad enough to cause decoherence within the integration time. See Chap. 7 for a full description of ionospheric effects and their impact on imaging observations.

Finally, the fourth of these states that the SIP must make an assumption about the primary calibration scheme; the use of non-standard calibration setups may lead to unpredictable results. Related to the issue of primary calibration, note that it is important to utilize smart demixing (Sect. 9.3.1) to obtain good quality amplitudes for the initial calibration.

9.3 SIP Strategy, Framework, and Components

In this section I describe the details of the SIP implementation. Much of the material is based on the description given by Heald et al. (2010), and updated where appropriate. For reference, a diagram illustrating the overall structure of the pipeline is shown in Fig. 9.1. The diagram will be fully explained in this section.

9.3.1 Preprocessing Pipeline

The initial processing stage for both calibrator and target observations makes use of NDPPP (for New Default Preprocessing Pipeline; refer to Chap. 4), which performs a number of low-level tasks including flagging and data compression. For our flagging strategy, we make use of the software developed by Offringa et al.

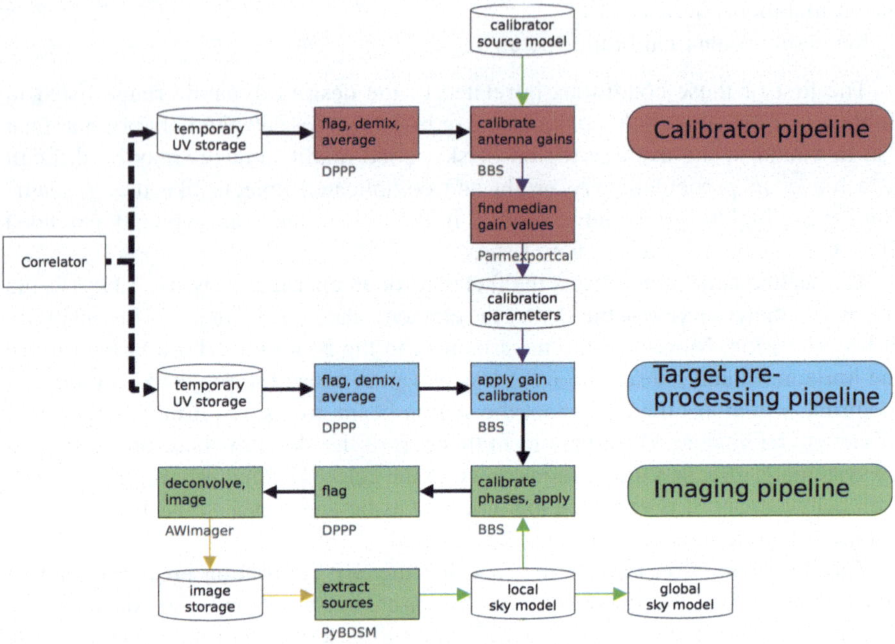

Fig. 9.1 Pipeline schematic

(2010), which has proven to work very well for LOFAR (as well as for other radio interferometers). There are two separate strategies (pre-defined sets of parameters) that are optimized for use with LBA and HBA observations. The LOFAR pipeline uses the appropriate strategy based on the observation.

LOFAR's location in the Netherlands means that it is embedded in an environment of high radio frequency interference (RFI) density in the local low-frequency radio spectrum. Indeed, the raw data show clearly that the spectrum contains numerous contributions from interfering sources. For detailed analysis, see the investigation by Offringa et al. (2013). Most interference is narrowband at these low frequencies, and is often sporadic as well. With this in mind, LOFAR was designed to provide extremely high frequency and time resolution during normal interferometric operations. The default frequency resolution is 3 kHz (each 195 kHz subband is made up of 64 channels), and typical integration times are either 1 s (in the LBA) or 2 s (in the HBA). The first flagging operation takes place on the full resolution data.

To illustrate the appearance of a typical LOFAR observation before and after flagging, we present in Fig. 9.2 a dataset before and after processing with NDPPP. Some regions of the spectrum have a much higher RFI occupancy than others, and these have been identified (e.g., Offringa et al. 2013) and are typically avoided in routine observations.

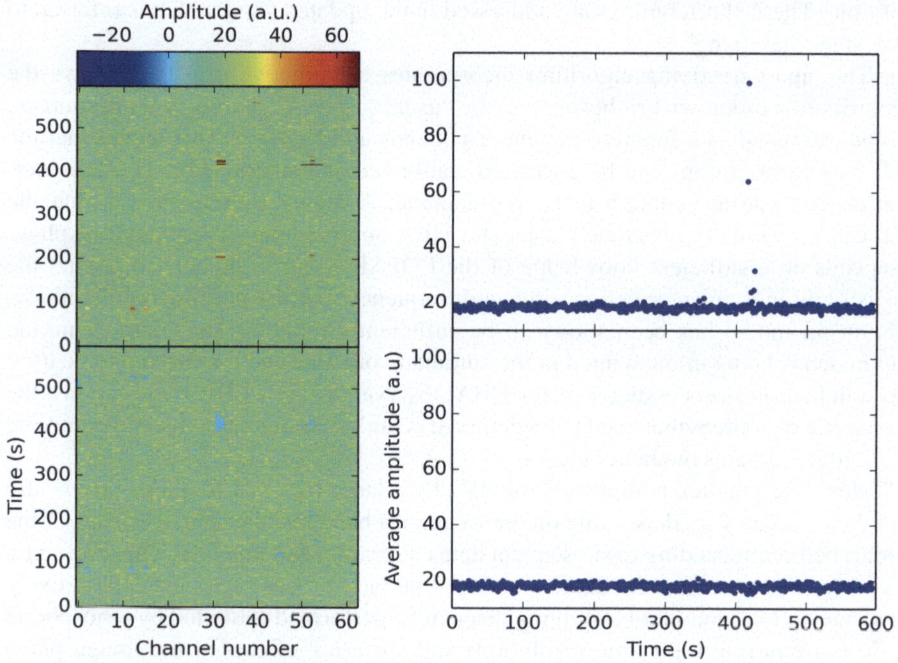

Fig. 9.2 A typical LOFAR HBA imaging dataset before and after flagging with NDPPP. On the left, the two plots show visibility amplitude in colorscale as a function of channel number and time, at the full spectral and time resolution of a normal imaging observation. Spikes of sporadic and narrowband RFI are visible in dark red in the top panel. The data are from a short calibrator scan on 3C295, are the equivalent of Stokes I (in other words are the sum of the XX and YY amplitudes), and represent a single subband at 149 MHz on the baseline between the two HBA ears of CS302. The top panel is the raw data from the correlator, and the bottom panel shows the result of running the `aoflagger` step in NDPPP using default settings. Flagged values are plotted here as zeros and therefore appear as light blue areas. The RFI spikes have clearly been removed. The plots on the right-hand side show the same visibility amplitudes, before (top) and after (bottom) flagging, after averaging over frequency. Again, the effect of the flagging is plainly visible through the removal of the spike of visibility amplitude at $t \sim 420$ s

After flagging, demixing is performed. This is a procedure aimed at removing far off-axis A-team sources from the visibilities. The technique itself is described by van der Tol et al. (2007) and is summarized in Chap. 4. In brief, the scheme aims to isolate, calibrate and subtract the contribution to the visibility function from bright sources far from the phase center of the observation. The same removal of off-axis flux can be achieved through directional calibration (e.g. with SageCal or BBS), but the benefit of demixing is that it can usually be performed in a blind fashion: i.e. with no prior detailed knowledge of the sources in the main field of view. Shortcomings include the fact that it subtracts off-axis flux from the directions being demixed at all times and on all baselines; and that the sources to demix must be specified ahead of time. These shortcomings are addressed in the updated version that is referred to as "smart demixing".

The smart demixing algorithm incorporates a predictive step that allows the contribution of known bright sources (the "usual suspects": mainly A-team sources) to be estimated as a function of time, frequency and baseline. This means that the off-axis contributions can be estimated, calibrated and subtracted only when their influence is strong enough to be problematic. A happy consequence is that the procedure normally runs much faster (for HBA observations). The prediction phase depends on a sufficient knowledge of the LOFAR beam to be able to predict the sensitivity in a given direction, time and frequency. For the purposes of demixing, the beam model has been shown to be sufficiently reliable. The smart demixing approach is being implemented in the automatic pipeline and is expected to be used as a default pre-processing step for HBA observations. For LBA observations, the influence of sources that need to be demixed is far less sporadic, so the old demixing procedure remains the better choice.

Next, the pipeline compresses the data by a large factor in frequency (typically 16, but this can vary depending on the frequency band and the characteristics of the field) before proceeding to subsequent data calibration and imaging. This averaging is performed to reduce the data volume so that later processing is not prohibitively expensive. The amount of averaging that can be performed is limited by ionospheric effects (requiring high time resolution) and smearing effects in the image plane (requiring high frequency resolution, and to a lesser degree high time resolution). For a typical HBA observation, the reduction in data volume is approximately an order of magnitude (averaged and calibrated data with respect to raw data from the correlator). The compression factor is not as simple as the averaging interval because it is offset by added data columns and metadata. As an example, a single subband of a 10-min calibrator observation may be 2.3 GByte in raw data, and 270 MByte after flagging, averaging (16$\times$ in frequency and 2$\times$ in time) and calibration. To first order these values scale with time, so a single subband from an 8 h target observation might require 110 GByte in raw data and 13 GByte in averaged and calibrated data. Note that HBA observations are typically considerably larger than LBA observations, since the core HBA stations are split into two antenna fields that are (optionally) correlated as separate stations. Thus, the data volume of a core-station HBA observation is about four times larger than a core-station LBA observation, all else being equal.

9.3.2 Calibration and Imaging Pipelines

The initial stages of the SIP are based on procedures that were defined and tested during early commissioning activity related to the LOFAR Multifrequency Snapshot Sky Survey (MSSS; Heald et al. 2015). The initial MSSS calibration and imaging effort was aimed at producing shallow, low resolution survey images with angular resolution $\approx 2'$ and dynamic range $\lesssim 1000$. In this scheme, the data are first phase calibrated using an initial model drawn from existing radio surveys, which is followed by an imaging step and mosaicing of adjacent fields.

9.3.2.1 Calibration Pipeline

Primary calibration is used to set the flux scale and to correct for the overall sensitivity variations. In the SIP this primary calibration takes place through the observation of one or more of a small number of suitable sources, together with the sky models defined by Scaife and Heald (2012). The sources in that set of primary calibrators are all compact and bright at LOFAR frequencies. Calibrator observations are either performed just before and/or after a target observation, or are observed simultaneously with the target. Usually the former is the case in HBA observations and the latter for LBA observations, where the analog beam does not limit the sensitivity in arbitrary directions on the sky. In both cases, the calibrator observation has the same frequency coverage as the target observation so that the calibrator solutions can be mapped one-to-one with the target data.

Calibrator observations are used to obtain station gains. The calibration step has traditionally been performed using BBS (Blackboard Selfcal), and this is what is currently implemented in the SIP. However, the calibration is being transitioned to NDPPP for efficiency reasons. Both calibration packages are based on the Hamaker-Bregman-Sault Radio Interferometer Measurement Equation (RIME; Hamaker et al. 1996; Smirnov 2011), which expresses the instrumental response to incoming electromagnetic radiation within the framework of a matrix formalism. Here, the various instrumental effects are identified, their effect on the signal is characterized in full polarization, and are quantified and parameterized as separate Jones matrices. Each of these terms may depend on different dimensions: frequency (e.g. the bandpass); time (e.g. the station gains); or direction (e.g. the station beam). Because it is based on the general form of the RIME, BBS can natively handle difficult problems such as direction dependent effects and full polarization calibration.

In Fig. 9.3, we illustrate station gain phases solved for by BBS on a calibrator source (here, 3C 295). A separate gain solution was obtained for every 2-s integration in the averaged data set. Each subband is calibrated separately. The LOFAR beam model was incorporated in the RIME, so the gains reflect amplitude and phase variations at each station in addition to predicted beam effects. Only two stations are shown here, but the others show the same typical behavior. The gain phase variations relative to the reference station (CS002HBA0 on the superterp) are larger and typically more rapid for increasingly distant stations, as expected.

The amplitude of the gains can be seen to be very stable in these calibrator observations. The rms variation with respect to the median value is typically 1.5% (as quantified in this particular observation). Before application to the target data, the gain amplitude solutions are median filtered to reject outliers. Primary calibration is completed by applying the resulting amplitude solution to the target field. In case of simultaneous calibrator and target observations (usually LBA), the median-filtered amplitudes can be applied directly to the target data including the time dependence.

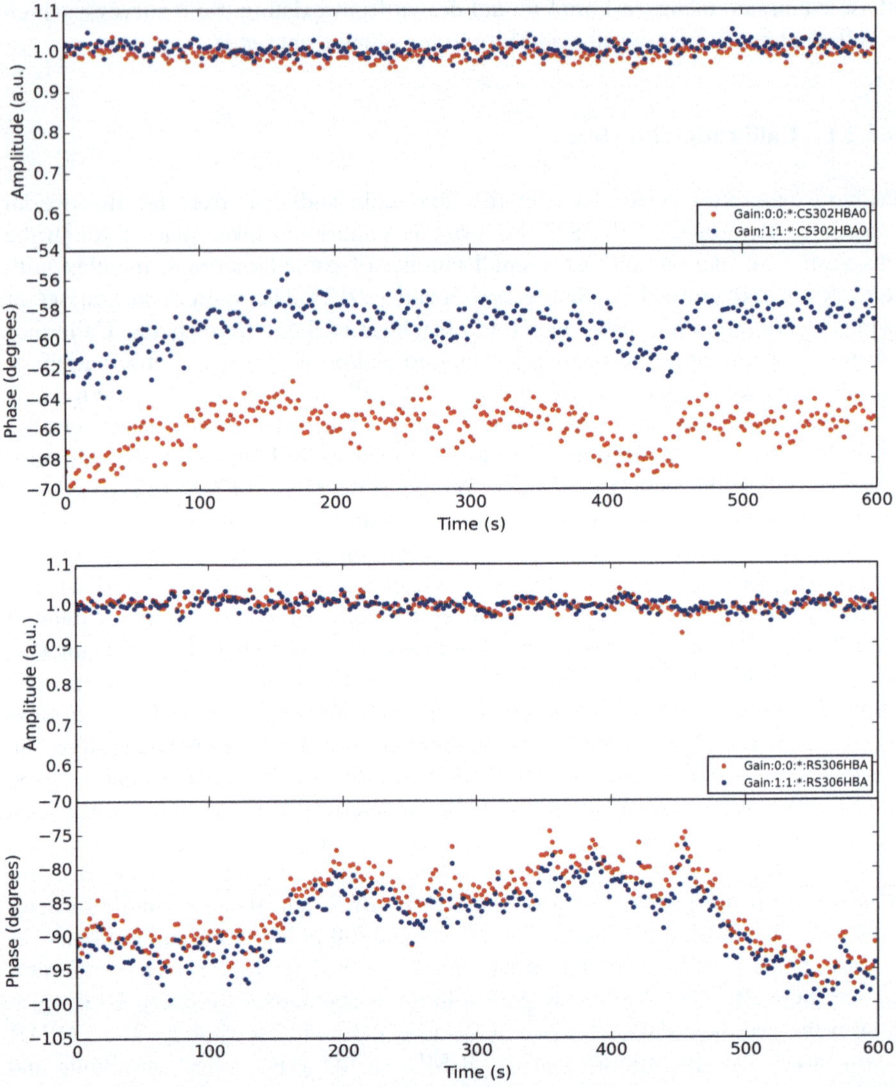

Fig. 9.3 Examples of station gain amplitudes and phases as determined by BBS. The reference station for the phases is CS002HBA0 on the superterp. The vertical axis shows the amplitude in arbitrary units, and the phase in degrees. In both cases, the red points show `Gain:0:0` (equivalent to the X gain) and the blue points show `Gain:1:1` (Y gain). The horizontal axis is the time in seconds. The amplitudes and phases are shown for CS302HBA0 (top) and RS306HBA (bottom). Note that the phases for the former generally vary more slowly and also show smaller peak-to-trough variation than the latter, which is located farther from the reference station. (CS302 is 1.8 km from the center of the superterp, whereas RS306 is 8.8 km away.) Note too that the amplitude variations in all cases are very small (in the case of both stations, the rms variation is 1.4–1.6% with respect to the median values). The results for these stations are representative of the quality of the gains derived for a typical HBA calibrator observation

When the calibrator is a separate observation (as is often the case in HBA), the median gain amplitude per station (and frequency) is applied to the target data as a time-independent quantity. Experience has shown that the gain amplitudes are stable over the course of several hours, making this strategy possible.

Following primary calibration, secondary phase calibration is performed within the target field of interest. A critical input to BBS (or NDPPP) is the sky model which is used to predict the visibilities. The SIP automatically constructs an initial sky model, currently based on cataloged values from the VLA Low-frequency Sky Survey redux (VLSSr; Cohen et al. 2007; Lane et al. 2014), the Westerbork Northern Sky Survey (WENSS; Rengelink et al. 1997), and the NRAO VLA Sky Survey (NVSS; Condon et al. 1998). This is the "Mark-0" LOFAR GSM; the "Mark-1" LOFAR GSM continues to be generated by the ongoing Multifrequency Snapshot Sky Survey (MSSS; Heald et al. 2015). MSSS will provide a considerably higher areal density of sources than the VLSSr catalog, and more importantly will ultimately include spectral information from 30 MHz up to 160 MHz. Recently, the TIFR GMRT Sky Survey Alternative Data Release (TGSS-ADR1; Intema et al. 2017) became available and is also extremely useful to generate sky models for HBA observations.

The secondary calibration is performed using concatenated subband groups (typically groups of 10 contiguous subbands, totalling 2 MHz bandwidth per group). Within each group a frequency-independent solution is obtained. The reason for doing this is to maximize the signal-to-noise in the resulting phase solutions. An example of this kind of output is shown in Fig. 9.4. As was also seen in Fig. 9.3, the phase variations are larger and more rapid for stations more distant from the reference station (here, CS002HBA0 on the superterp). Note too that the gain phases become more noisy for distant stations (particularly RS508HBA in this example), because the sky model becomes inaccurate at high resolution and the importance of direction-dependent ionospheric effects is becoming increasingly dominant. Dealing with these issues is the subject of the advanced pipeline processing that is now under development (see Sect. 9.5).

9.3.2.2 Imaging

Imaging of LOFAR data is performed using the AWImager (Tasse et al. 2013), which as described in Chap. 8 is a custom-built variant of the Common Astronomy Software Applications (CASA; Jaeger 2008) imager including both A-projection (Bhatnagar et al. 2008) and w-projection (Cornwell et al. 2008). The A-term includes the full LOFAR beam model at a frequency and time resolution that can be selected by the user. The SIP chooses these resolution values by default to be one value per subband and per 5 min time interval. The AWImager is multithreaded and is a component that is currently under intense development as described in Sect. 9.5. Imaging with LOFAR and the required software is described in great detail in Chap. 8.

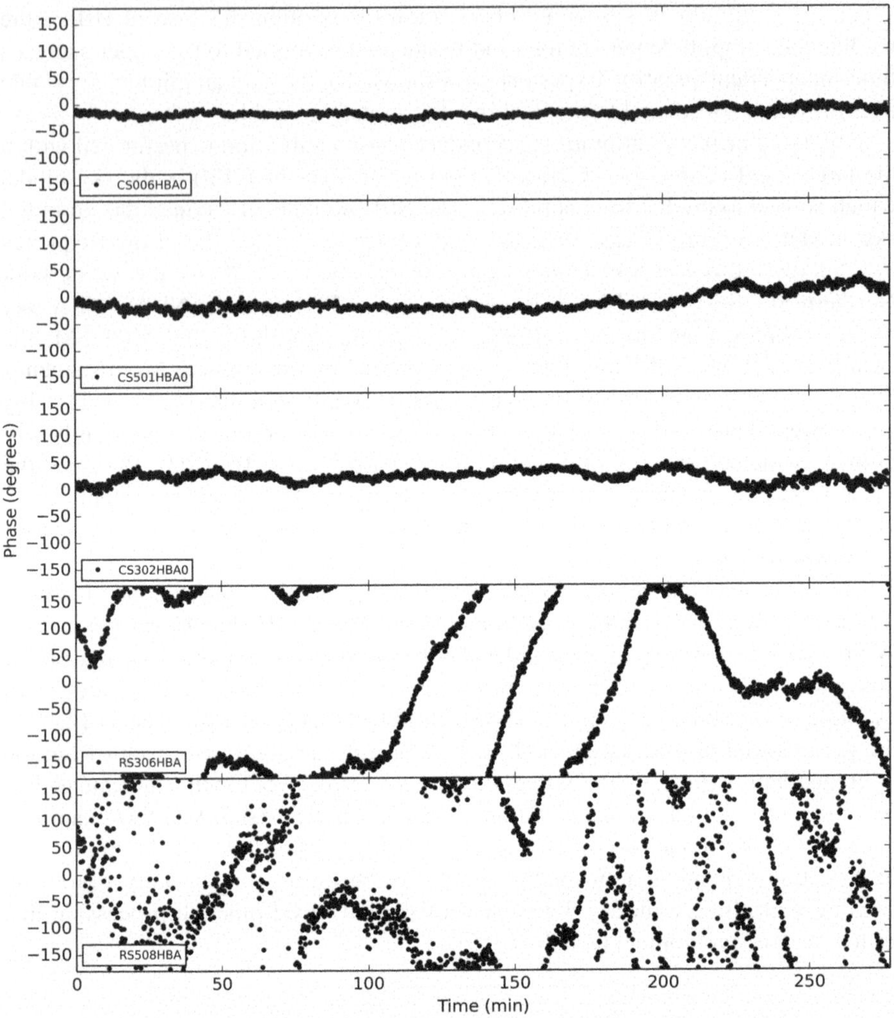

Fig. 9.4 Examples of station gain phases as determined by BBS, using a 10-subband group (2 MHz bandwidth). The subband group illustrated here spans 149–151 MHz. Only the `Gain:0:0` (X gain) is shown; the `Gain:1:1` (Y gain) behavior is essentially identical. The phases are all referenced to CS002HBA0 on the superterp. Only five representative stations are shown (one per panel). The station names are given in each panel. The distances from the reference station are as follows. In order of increasing distance from the reference station (top to bottom), they are CS006HBA0: 190 m (also on the superterp); CS501HBA0: 1.1 km; CS302HBA0: 1.8 km; RS306HBA: 8.8 km; RS508HBA: 29.5 km

Using the data for which the calibration in Fig. 9.4 was derived, an example output image is shown in Fig. 9.5. Here a moderate-resolution image is shown together with the average primary beam. Note that this field is located at a declination near the celestial equator, so the beam is extended due to foreshortening

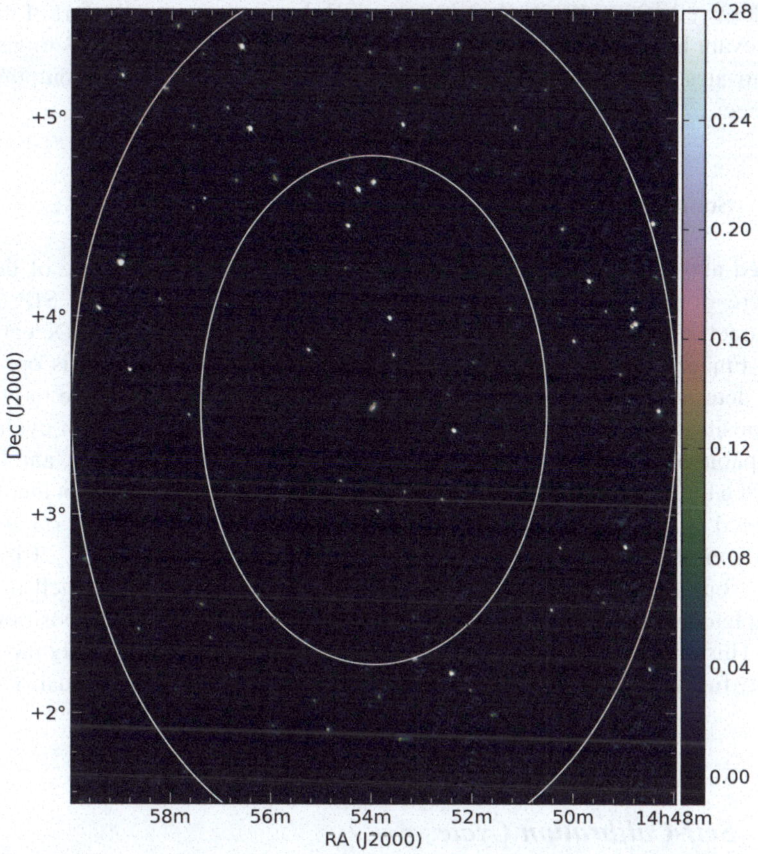

Fig. 9.5 Intermediate-resolution image produced from data calibrated as shown in Fig. 9.4. The image was made using the AWImager, with Briggs weighting (robust=−0.5), a maximum baseline length of 8 km, and an inner cut in the uv plane of 200 m. The resulting beamsize is $53'' \times 38''$ (PA = 25°) and the rms noise level is approximately 6 mJy beam^{-1} near the center of the image. The average primary beam (provided as a standard output of AWImager) is displayed with white contours: the inner contour is the 80% level, and the outer contour is the 50% level

of the station size (relevant when observing toward low elevation). At higher declination (elevation), the beam pattern would be more circular.

In cases where adjacent overlapping fields have been observed, calibrated and imaged, image mosaics can be produced using a hand-crafted script (`mos.py`) that was produced for MSSS purposes. An updated version of the technique has been employed for the LoTSS (Shimwell et al. 2017) survey, and other tools such as `SWARP`[1] can also be used. The script makes use of the same inverse-variance weighting scheme that has long been successfully used for other radio telescopes

[1] https://www.astromatic.net/software/swarp.

(e.g., Cornwell 1988). While this mosaic capability is not currently part of the SIP, it is relevant here because the sensitivity pattern that is used in MSSS to generate large mosaics is the primary beam map that is produced as a standard output of the AWImager.

9.3.2.3 Source Finding

As noted above a crucial part of successful calibration, and therefore of the SIP, is the creation of a reliable source model. The software used in the SIP to find and characterize sources in LOFAR images is called the Python Blob Detector and Source Finder (PyBDSF[2]; Mohan and Rafferty 2015). The software is optimized for the detection of both unresolved and extended sources, and is able to produce decompositions of sources for use in BBS to continue the calibration cycle. It is also capable of recognizing regions of images suffering from artifacts, and locally raising the thresholds for identification of sources so that artifacts are not incorrectly interpreted as astronomical objects. PyBDSF has recently been tested on artificial data in comparison with several other source finders (Hopkins et al. 2015), and in that "source finding challenge" has been shown to perform very well in terms of completeness and reliability, as well as proper recovery of source positions and fluxes. This excellent performance is crucial so that BBS can accurately model the visibility function of the corresponding observation and produce high-quality station gains.

9.3.3 Self-Calibration Cycle

So far, the calibration (and subsequent imaging) described as part of the SIP has only been referenced to external *a priori* knowledge (primary calibrator models, and GSM information from existing radio surveys). Particularly in the case of the GSM model used for the phase calibration step in the Imaging Pipeline, this information is not sufficient to get the most out of the LOFAR data: the initial sky model is produced from observations made at low resolution and/or at different radio frequencies than the actual observations. Moreover, it has long been known that achieving the best data quality requires an iterative cycle that progressively updates both the sky model and the instrumental model. This aspect of the data reduction cycle is now available in the SIP.

A direction independent self-calibration cycle has recently been implemented as part of the SIP. This basic scheme performs a loop of phase calibration, imaging and model creation. The loop usually begins at low resolution and proceeds to higher resolution in a staged approach (each sequence of calibration, imaging and

[2]PyBDSF documentation: http://www.astron.nl/citt/pybdsm/.

model update is referred to as a "major cycle"). In practice, the progression of image resolution can be selected as a user input. Experience in the user community suggests that only a small number of iterations are required to progress from arcminute resolution to the best image quality that can be expected from a direction independent strategy. It has been found that the SIP self-calibration loop is sufficient to achieve reasonable image quality at the $\approx$10–20$''$ level, and often as good as 5$''$ in certain fields. Corresponding image noise levels are also reasonably close to the theoretical thermal noise limit, but image artifacts naturally arise in images produced at the highest resolution because the direction independent approach breaks down; a direction dependent strategy is required to obtain the highest quality images (see Sect. 9.5 for progress in that area).

9.4 Using the SIP

As an observer, there are several aspects to take into account before using the SIP. Most of the settings within the pipeline are properly configured and either cannot be changed or do not need to be changed. However, there are some aspects to consider and they are briefly discussed here.

Demixing (see Sect. 9.3.1 and Chap. 4) is an important pipeline step that needs to be considered. Since the output of the demixing step sets the quality of the data that will be used for all subsequent steps, it is well worth the effort to take particular care in the demixing configuration. If your processing strategy will make use of directional calibration including far off-axis (A-team) sources, then demixing can be disabled (In this case, however, you are probably not using the SIP.). Otherwise, the following considerations apply. For HBA observations, the smart demixing algorithm is ideal (when it is made available). Until then, the process for selecting sources to demix as described in Chap. 4 should be followed. For LBA observations, smart demixing is unlikely to be a wise choice since interfering sources will always be present. It is likely that Cas A and Cyg A should always be demixed; additional sources may also be needed, again based on the description in Chap. 4.

The averaging step also bears attention. Averaging in frequency is normally set to 16 (for observations with 64 channels per subband), such that the output subbands have 4 channels each. This high frequency resolution is often needed to minimize the effect of bandwidth smearing, and may also be needed for polarization analysis (see Chap. 10). Time averaging is commonly dictated more by the timescale of ionospheric effects than by time smearing, and is usually set in such a way as to provide time samples separated by $\approx$10 s.

For the imaging pipeline, the field of view needs to be considered. The nominal field of view of different configurations is given by van Haarlem et al. (2013), but note that sources will be visible well past the nominal HPBW of the station beam. Imaging an area at least twice as wide is advisable. Also keep in mind that the field of view will stretch in the north-south direction for observations at low declination.

In the self-calibration step, the sequence of image resolution for each major cycle is an important setting. The determination of a recommended sequence is still underway at the time of writing, but it is likely that a small number of major cycles, advancing from arcminute resolution to $\approx 10''$–$20''$ resolution in just 2–3 steps, will provide the most efficient path to a reasonable quality image. Retaining the visibilities is highly recommended—it is likely that the SIP can provide a good image for initial analysis, but it is also likely that an improved image will be required for scientific publication. The need for post-SIP processing will hopefully be reduced or even completely eliminated with the advent of the direction-dependent algorithm (facet calibration) now under development in the SIP framework, as described in the next section.

Last but certainly not least, it is essential to consider the runtime of the selected pipeline configuration. Aspects such as number of sources subtracted during demixing, and the chosen image size, will substantially affect how long the SIP runs. This quantity is typically characterized as the "P/O" ratio, or the ratio of processing (P) to observing (O) time. Each component of the pipeline has its own P/O ratio. For the most up-to-date information regarding pipeline performance and the best understanding of the relevant P/O ratios, the reader is referred to the Radio Observatory's online documentation:
http://www.astron.nl/radio-observatory/astronomers/array-configurations/3-telescope-parameters-and-array-configurations.

9.5 Limitations and Future Work

The current version of the SIP can be improved in a few fundamental ways. First, some of the steps can be sped up (e.g. imaging) so that more processing can be done within a fixed amount of time. Second, some of the steps can be better automated (e.g. demixing) and can use better algorithmic settings. Third, moving from direction-independent to direction-dependent processing is required in order to achieve the quality of output image that is required for science readiness. Enabling direction-dependent calibration is a key goal of the SIP development.

Current developments to extend the SIP are therefore focused on the following four key areas:

- Calibration efficiency
- Imaging capability
- Self-calibration enhancements
- Direction-dependent calibration

Each of these areas are now briefly described in turn.

Calibration Efficiency The solver that is incorporated in BBS uses a reliable algorithm that has been used for decades: the Levenberg-Marquardt (LM) least-squares solver. However, other solving algorithms are known to be more efficient in reaching the same quality solutions. For example, the SAGE algorithm (see Yatawatta et al. 2008; Kazemi et al. 2013) has proven to be very efficient in solving direction-dependent problems, and other optimized solver algorithms are now becoming available (e.g., employing Wirtinger derivatives; Smirnov and Tasse 2015). Even direction-independent algorithms can provide marked improvements with respect to the LM solver at the heart of BBS. One of these makes use of alternating direction implicit (ADI) methods; see Salvini and Wijnholds (2014). This algorithm has been implemented as a fast solver in the NDPPP program (step `gaincal`). Use of this solver in place of BBS can provide a much more efficient path to high-quality solutions, in specific circumstances (e.g., determination of primary calibrator solutions, or phase-only calibration). This solver will be used where possible in the SIP in order to make the calibration steps much more efficient.

Imaging Capability Often, the imaging steps are the most time-consuming for a calibration and imaging pipeline. This is particularly true in the case of the high-resolution, large field of view imaging jobs that are required to make use of standard LOFAR imaging observations. Recent SIP development has seen substantial effort into a reliable imaging framework (the AWImager) which not only makes use of all the recent deconvolution algorithms used in CASA (multiscale and broadband deconvolution options), but also allows for modular use of the gridding algorithm (see Chap. 8). In this way, the efficiency of the imaging step can be improved through the development of a fast gridding algorithm and subsequently "plugging it in" to the AWImager framework. Such an improved gridding algorithm is "Image Domain Gridding" (van der Tol et al., in prep) and will be incorporated in the SIP. It is anticipated that this new scheme will provide a dramatic speedup in the imaging steps.

Self-Calibration Enhancements The self-calibration algorithm used in the SIP is based on user experience (not only from LOFAR, but from calibrating and imaging data from a broad range of radio interferometers). Particular details can still be enhanced, and these improvements are being implemented in the SIP. For example a key enhancement is the identification of the optimum progression of image resolution during the sequence of major cycles. As highlighted earlier, a key aspect of the self-calibration cycle is ensuring the highest quality of sky models during the calibration process; therefore variations on the sky model development are being explored for use in the SIP (e.g., using the PyBDSF output not to directly generate a sky model, but instead to create a mask for `CLEAN` deconvolution, so that the `CLEAN` components can be used as the sky model as is typically done with other radio interferometers).

Direction-Dependent Calibration The primary enhancement of the SIP will be the transition from direction-independent to direction-dependent calibration. This development incorporates improvements in all areas described above, together with other new features such as user-friendly interfaces to generate and manage instrument models and sky models (`LoSoTo` by Francesco de Gasperin and `LSMTool` by David Rafferty, respectively). All of this progress is being integrated in order to build an enhanced SIP system using the "facet calibration" algorithm developed by Reinout van Weeren and collaborators. The scheme is fully described by van Weeren et al. (2016) and Williams et al. (2016). It breaks up the field of view of a LOFAR observation into a manageable number of facets, using a Voronoi tessellation. The number of facets is chosen based on the availability of sufficiently bright calibrator sources in the field of view, and the desire to keep the facet size small (at least within the central part of the field of view). Within each facet, a particular self-calibration sequence is performed to build up an excellent gain solution using the calibrator source. The gains include a fast ($\approx$10 s) phase component to capture ionospheric variation, and a slow ($\sim$10 min) amplitude component to capture beam variations. The dimensions of the gains are chosen in such a way as to minimize the number of free parameters utilized throughout the entire procedure. Each facet is dealt with individually, which is made possible by subtracting all other facets from the visibilities in advance. In that way this scheme is a variant of the well-known peeling scheme (e.g., Noordam 2004; Intema et al. 2009) and has therefore been referred to as "Extreme Peeling". After all facets are calibrated and imaged, they are all mosaiced together at the end to generate an image of the full field of view. This procedure is being implemented in an enhanced pipeline framework that enables rapid development and straightforward realization of complicated processing schemes. Looking to the future, a variant of this process that uses the per-facet solutions to develop an ionospheric TEC or phase screen (see Chap. 7) and applies that screen in the AWImager—similar to the Source Peeling and Atmospheric Modeling (SPAM) technique (Intema et al. 2009)—may also be developed.

As a final remark, it is essential to note the key contributions of researchers in the LOFAR community. Throughout the development of the SIP and on an ongoing basis, the user community has provided fundamental contributions, ideas, and feedback to the processing software and algorithms. Without these contributions LOFAR calibration and imaging would not have come as far as it has.

Acknowledgements I would like to thank Tammo Jan Dijkema for the pipeline figure, and the entire "Calibration and Imaging Tiger Team" (CITT) based at ASTRON and elsewhere for their tireless efforts to expand the SIP into the arena of fast and efficient direction-dependent calibration. At the time that this description was written, the CITT included Tammo Jan Dijkema, Stefan Fröhlich, GH, Emanuela Orrú, David Rafferty, Tim Shimwell, Bas van der Tol, Carmen Toribio, and Nicolas Vilchez.

References

Bhatnagar, S., Cornwell, T.J., Golap, K., Uson, J.M.: Astron. Astrophys. **487**, 419 (2008)
Cohen, A.S., Lane, W.M., Cotton, W.D., et al.: Astron. J. **134**, 1245 (2007)
Condon, J.J., Cotton, W.D., Greisen, E.W., et al.: Astron. J. **115**, 1693 (1998)
Cornwell, T.J.: Astron. Astrophys. **202**, 316 (1988)
Cornwell, T.J., Golap, K., Bhatnagar, S.: IEEE J. Sel. Top. Signal Process. **2**, 647 (2008)

Hamaker, J.P., Bregman, J.D., Sault, R.J.: Astron. Astrophys. Suppl. Ser. **117**, 137 (1996)
Heald, G., McKean, J., Pizzo, R., et al.: ArXiv e-prints (2010)
Heald, G.H., Pizzo, R.F., Orrú, E., et al.: Astron. Astrophys. **582**, A123 (2015)
Hopkins, A.M., Whiting, M.T., Seymour, N., et al.: Publ. Astron. Soc. Aust. **32**, e037 (2015)
Intema, H.T., van der Tol, S., Cotton, W.D., et al.: Astron. Astrophys. **501**, 1185 (2009)
Intema, H.T., Jagannathan, P., Mooley, K.P., Frail, D.A.: Astron. Astrophys. **598**, A78 (2017)
Jaeger, S.: In: Argyle, R.W., Bunclark, P.S., Lewis, J.R. (eds.) Astronomical Data Analysis Software and Systems XVII. Astronomical Society of the Pacific Conference Series, vol. 394, p. 623 (2008)
Kazemi, S., Yatawatta, S., Zaroubi, S.: Mon. Not. R. Astron. Soc. **434**, 3130 (2013)
Lane, W.M., Cotton, W.D., van Velzen, S., et al.: Mon. Not. R. Astron. Soc. **440**, 327 (2014)
Mohan, N., Rafferty, D.: PyBDSM: Python Blob Detection and Source Measurement, Astrophysics Source Code Library (2015)
Noordam, J.E.: In: Oschmann, J.M., Jr. (ed.) Ground-Based Telescopes. Society of Photo-Optical Instrumentation Engineers (SPIE) Conference Series, vol. 5489, pp. 817–825 (2004)
Offringa, A.R., de Bruyn, A.G., Biehl, M., et al.: Mon. Not. R. Astron. Soc. **405**, 155 (2010)
Offringa, A.R., de Bruyn, A.G., Zaroubi, S., et al.: Astron. Astrophys. **549**, A11 (2013)
Rengelink, R.B., Tang, Y., de Bruyn, A.G., et al.: Astron. Astrophys. Suppl. Ser. **124**, 259 (1997)
Salvini, S., Wijnholds, S.J.: Astron. Astrophys. **571**, A97 (2014)
Scaife, A.M.M., Heald, G.H.: Mon. Not. R. Astron. Soc. **423**, L30 (2012)
Shimwell, T.W., Röttgering, H.J.A., Best, P.N., et al.: Astron. Astrophys. **598**, A104 (2017)
Smirnov, O.M.: Astron. Astrophys. **527**, A106 (2011)
Smirnov, O.M., Tasse, C.: Mon. Not. R. Astron. Soc. **449**, 2668 (2015)
Tasse, C., van der Tol, S., van Zwieten, J., van Diepen, G., Bhatnagar, S.: Astron. Astrophys. **553**, A105 (2013)
van der Tol, S., Jeffs, B.D., van der Veen, A.-J.: IEEE Trans. Signal Process. **55**, 4497 (2007)
van Haarlem, M.P., Wise, M.W., Gunst, A.W., et al.: Astron. Astrophys. **556**, A2 (2013)
van Weeren, R.J., Williams, W.L., Hardcastle, M.J., et al.: Astrophys. J. Suppl. Ser. **223**, 2 (2016)
Williams, W.L., van Weeren, R.J., Röttgering, H.J.A., et al.: Mon. Not. R. Astron. Soc. **460**, 2385 (2016)
Yatawatta, S., Zaroubi, S., de Bruyn, G., Koopmans, L., Noordam, J.: ArXiv e-prints (2008)

Part III
Advanced Topics in LOFAR Data Processing

Chapter 10
Polarization Imaging with LOFAR

Michiel A. Brentjens

Abstract Polarimetry with LOFAR is somewhat different from polarimetry with, for example, the VLA or WSRT. At LOFAR frequencies, Faraday rotation is much stronger than at for example L band. The ionosphere in particular is a big nuisance. Moreover, the polarized response of (a group of) phased array stations is very different from the response of (a group of) parabolic dish antennas. This lecture focuses mainly on the physics of low frequency polarimetry, and how it affects observation planning, data processing, and imaging.

10.1 Introduction

Radio antennas are fundamentally polarized detectors. Even for making total intensity images, it is essential to understand polarimetry. There are already many texts treating the basics of polarization of electromagnetic waves (e.g. Born and Wolf 1999) and its significance for radio interferometry using narrow-beam dish antennas (e.g. Cotton 1999; Thompson et al. 2001), hence this chapter merely summarizes the basic concepts.

Stationary phased array antenna stations have very different polarimetric properties than parabolic dishes that track the sky. The polarimetric response varies on scales of the entire sky, instead of on scales of the antenna station's beam. In fact, if a station is well calibrated, the analogue tile beams and digital station beams should not introduce any additional polarimetric distortion.

Faraday rotation, the process that rotates the plane of polarization of linearly polarized radiation propagating through a magnetized plasma, is proportional to the square of the wavelength. It is therefore a factor ~100–1000 more severe at LOFAR frequencies than at L-band. Besides bandwidth depolarization due to interstellar Faraday rotation, one also has to worry about time-dependent effects due

M. A. Brentjens (✉)
ASTRON, Dwingeloo, The Netherlands
e-mail: brentjens@astron.nl

G. Heald et al. (eds.), *Low Frequency Radio Astronomy and the LOFAR Observatory*, Astrophysics and Space Science Library 426,
https://doi.org/10.1007/978-3-319-23434-2_10

to the Earth's ionosphere. They not only affect polarimetry, but also total intensity imaging, particularly on baselines longer than a few tens of km.

10.2 Polarized Electromagnetic Waves

In the fully general case, the tip of the electric vector of an electromagnetic wave traces an ellipse as a function of time, in a plane perpendicular to its propagation direction, and at a fixed point in space. Figure 10.1 shows one such example when viewing towards the *source* of the radiation. The shape of this ellipse contains all information about the polarization state of a monochromatic wave. There are two convenient ways to describe the ellipse: relative to a Cartesian basis, and relative to a circular basis. These bases are of course closely related to the ways linear and circular feeds detect radio waves.

In the Cartesian (or linear) basis,

$$\mathbf{E} = E_{\mathrm{x}}\hat{\mathbf{e}}_{\mathrm{x}} + E_{\mathrm{y}}\hat{\mathbf{e}}_{\mathrm{y}}, \tag{10.1}$$

where $\mathbf{E}$ is the electric field vector, $\hat{\mathbf{e}}_x = (1, 0)^{\mathrm{T}}$ and $\hat{\mathbf{e}}_y = (0, 1)^{\mathrm{T}}$ are the basis vectors, and E_x and E_y are the rapidly oscillating amplitudes in the x and y directions:

$$E_{\mathrm{x}} = A_{\mathrm{x}} \cos(2\pi \nu t + \delta_{\mathrm{x}}) \tag{10.2}$$

$$E_{\mathrm{y}} = A_{\mathrm{y}} \cos(2\pi \nu t + \delta_{\mathrm{y}}). \tag{10.3}$$

Here, A_{x} and A_{y} are the time-independent x- and y-amplitudes. $\delta_{\mathrm{xy}} = \delta_{\mathrm{y}} - \delta_{\mathrm{x}}$ is the xy phase difference: a measure of the wave's ellipticity. If $\delta_{\mathrm{xy}} > 0$ one has clockwise rotation or left *elliptical* polarization (LEP), if 0 *linear* polarization, and if

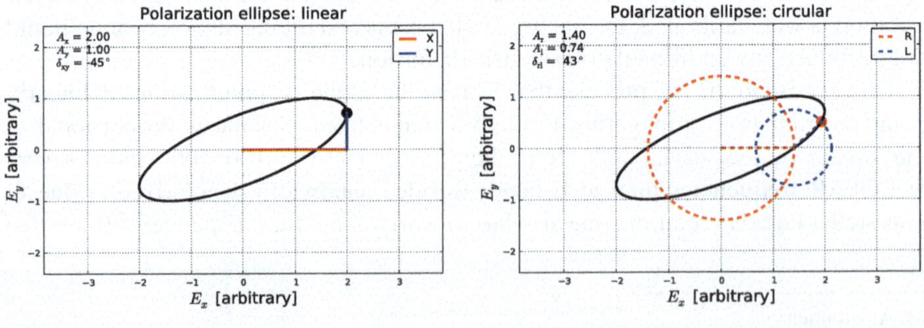

Fig. 10.1 Polarization ellipse in Cartesian (left) and circular basis (right). The ellipse traces the tip of the electric vector as a function of time, in a plane that is stationary in space, looking towards the radiation's source

< 0 counter clockwise rotation or right elliptical polarization (REP). If $\delta_{xy} = \pm 90°$, one has left- or right *circular* polarization.

The ellipse can also be described in terms of circularly rotating basis vectors:

$$\mathbf{E} = A_r \hat{\mathbf{e}}_r + A_l \hat{\mathbf{e}}_l, \tag{10.4}$$

where

$$\hat{\mathbf{e}}_r = \begin{pmatrix} \cos(2\pi \nu t + \delta_r) \\ \sin(2\pi \nu t + \delta_r) \end{pmatrix} \tag{10.5}$$

$$\hat{\mathbf{e}}_l = \begin{pmatrix} \cos(2\pi \nu t + \delta_l) \\ -\sin(2\pi \nu t + \delta_l) \end{pmatrix} \tag{10.6}$$

and A_r and A_l are the right- and left circular amplitudes, respectively. The sum of these amplitudes is the semi-major axis of the ellipse, while the absolute value of their difference equals the semi-minor axis. The rl phase difference $\delta_{rl} = \delta_l - \delta_r$ is a measure of the ellipse's orientation. More precisely, the position angle is $-\frac{1}{2}\delta_{rl}$. Therefore, if $\delta_{rl} > 0$, the major axis is rotated clockwise with respect to the x-axis, if 0 it is aligned to the x-axis, and if < 0 it is rotated counter clockwise.

10.3 Stokes Parameters

As demonstrated in the previous section, three parameters suffice to fully describe a monochromatic wave's polarization state. One more is needed to also account for partially polarized quasi-monochromatic radiation. It would be convenient if these parameters had the same units. Stokes (1852) for the first time described four parameters which all have units of power, and which describe the polarization state of both fully and partially polarized radiation. These Stokes parameters were introduced to astronomy by Chandrasekhar (1946).[1]

Using the conventions from the previous section, the Stokes parameters of a strictly monochromatic wave are

$$I = A_x^2 + A_y^2 \qquad\qquad I = A_r^2 + A_l^2 \tag{10.7}$$

$$Q = A_x^2 - A_y^2 \qquad\qquad Q = 2A_r A_l \cos \delta_{rl} \tag{10.8}$$

[1]The reader is encouraged to read these papers. Please note that these works did not use the same naming convention as is currently used. In fact, Stokes used A, B, C, and D, and Chandrasekhar I_r, I_l, U, and V, where I_r and I_l represent *rectilinear* (y) and *linear* (x) intensities, not right and left circular.

$$U = 2A_x A_y \cos\delta_{xy} \qquad U = -2A_r A_l \sin\delta_{rl} \tag{10.9}$$

$$V = -2A_x A_y \sin\delta_{xy} \qquad V = A_r^2 - A_l^2. \tag{10.10}$$

Here, I represents total intensity, Q and U describe linear polarization, and V circular. Because a strictly monochromatic wave is by definition fully polarized,

$$I^2 = Q^2 + U^2 + V^2. \tag{10.11}$$

Of course, monochromatic radiation does not exist. In radio interferometry one usually uses quasi-monochromatic radiation, which has a finite bandwidth $\Delta\nu$, and for which, after correlation or beam forming, the Stokes parameters are averaged over a time span $\tau \gg \Delta\nu^{-1}$. The average quasi-monochromatic Stokes parameters are then

$$I = \langle A_x^2\rangle + \langle A_y^2\rangle \qquad I = \langle A_r^2\rangle + \langle A_l^2\rangle \tag{10.12}$$

$$Q = \langle A_x^2\rangle - \langle A_y^2\rangle \qquad Q = \langle 2A_r A_l \cos\delta_{rl}\rangle \tag{10.13}$$

$$U = \langle 2A_x A_y \cos\delta_{xy}\rangle \qquad U = \langle -2A_r A_l \sin\delta_{rl}\rangle \tag{10.14}$$

$$V = \langle -2A_x A_y \sin\delta_{xy}\rangle \qquad V = \langle A_r^2\rangle - \langle A_l^2\rangle, \tag{10.15}$$

where $\langle\cdot\rangle$ denotes averaging in time and frequency. In this case,

$$I^2 \geq Q^2 + U^2 + V^2. \tag{10.16}$$

We also define the fractional linear polarization $p = \sqrt{Q^2 + U^2}/I \leq 1$ and fractional circular polarization $v = \|V\|/I \leq 1$.

Figure 10.2 shows what the Stokes parameter conventions adopted by the International Astronomical Union (IAU) look like on the sky. Polarization angle is measured north through east. Right Circular Polarization (RCP) and Left Circular Polarization (LCP) indicate the movement of the electric field vector tip in a plane that is at a fixed location in space, when looking towards the radiation's source. Always use this convention when writing about polarimetric results. Because

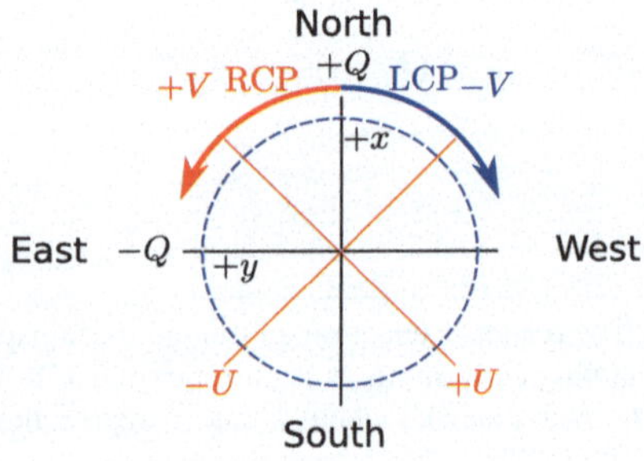

Fig. 10.2 Polarization nomenclature and sign conventions as defined in IAU (1973) and explained in Hamaker and Bregman (1996)

previously different papers have adopted different conventions—and some still do—one should exercise extreme caution when interpreting polarimetric papers, and one should never *assume* that this convention is used.

10.4 Stokes Visibilities

Synthesis radio telescopes do not directly measure images of the sky. Instead, they record the Fourier transforms of those images: the visibilities. Because the Stokes parameters have dimensions of brightness, we can define visibilities for all of them, not just Stokes I:

$$\mathscr{I}(u, v) = \mathscr{F}^{+}(I(l, m)) \tag{10.17}$$

$$\mathscr{Q}(u, v) = \mathscr{F}^{+}(Q(l, m)) \tag{10.18}$$

$$\mathscr{U}(u, v) = \mathscr{F}^{+}(U(l, m)) \tag{10.19}$$

$$\mathscr{V}(u, v) = \mathscr{F}^{+}(V(l, m)). \tag{10.20}$$

where in the 2D case,

$$\mathscr{F}^{+}(f) = \int_{lm} f \mathrm{e}^{+2\pi \mathrm{i}\nu(ul+vm)/c} \mathrm{d}l \mathrm{d}m. \tag{10.21}$$

The quantities $\mathscr{I}$, $\mathscr{Q}$, $\mathscr{U}$, and $\mathscr{V}$ are the Stokes visibilities. We will now derive how to compute them from the cross correlations produced by the digital back end (Fig. 10.3).

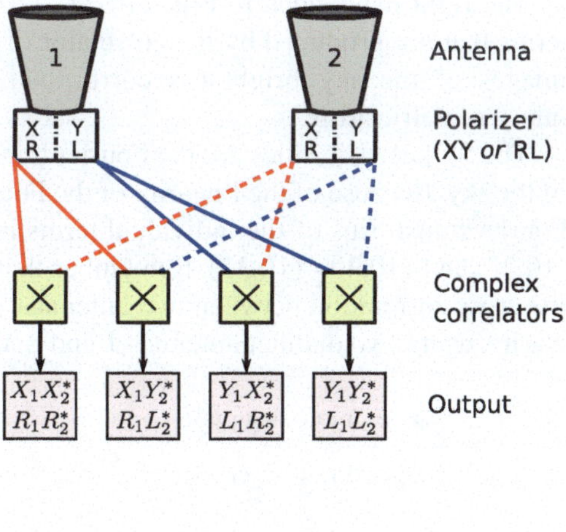

Fig. 10.3 A full polarization complex cross correlator as is used in every correlating radio telescope. In case of a single dish, the antenna 2 signal is actually a copy of antenna 1's signal

When rewriting E_x, E_y as the real part of complex quantities

$$E_{\mathrm{x}} = \Re\left\{A_{\mathrm{x}}\mathrm{e}^{2\pi\mathrm{i}\nu t}\right\} \tag{10.22}$$

$$E_{\mathrm{y}} = \Re\left\{A_{\mathrm{y}}\mathrm{e}^{\mathrm{i}\delta_{\mathrm{xy}}}\mathrm{e}^{2\pi\mathrm{i}\nu t}\right\}, \tag{10.23}$$

the Stokes parameters can be rewritten as

$$I = \langle A_{\mathrm{x}}^2\rangle + \langle A_{\mathrm{y}}^2\rangle \qquad = \langle E_{\mathrm{x}}E_{\mathrm{x}}^*\rangle + \langle E_{\mathrm{y}}E_{\mathrm{y}}^*\rangle \tag{10.24}$$

$$Q = \langle A_{\mathrm{x}}^2\rangle - \langle A_{\mathrm{y}}^2\rangle \qquad = \langle E_{\mathrm{x}}E_{\mathrm{x}}^*\rangle - \langle E_{\mathrm{y}}E_{\mathrm{y}}^*\rangle \tag{10.25}$$

$$U = \langle 2A_{\mathrm{x}}A_{\mathrm{y}}\cos\delta_{\mathrm{xy}}\rangle \qquad = \langle E_{\mathrm{x}}E_{\mathrm{y}}^*\rangle + \langle E_{\mathrm{y}}E_{\mathrm{x}}^*\rangle \tag{10.26}$$

$$V = \langle -2A_{\mathrm{x}}A_{\mathrm{y}}\sin\delta_{\mathrm{xy}}\rangle \qquad = -\mathrm{i}\left(\langle E_{\mathrm{x}}E_{\mathrm{y}}^*\rangle - \langle E_{\mathrm{y}}E_{\mathrm{x}}^*\rangle\right). \tag{10.27}$$

The same can be done in the circular basis:

$$E_{\mathrm{r}} = \Re\left\{A_{\mathrm{r}}\mathrm{e}^{2\pi\mathrm{i}\nu t}\right\} \tag{10.28}$$

$$E_{\mathrm{l}} = \Re\left\{A_{\mathrm{l}}\mathrm{e}^{-\mathrm{i}\delta_{\mathrm{rl}}}\mathrm{e}^{-2\pi\mathrm{i}\nu t}\right\} \tag{10.29}$$

$$I = \langle A_{\mathrm{r}}^2\rangle + \langle A_{\mathrm{l}}^2\rangle \qquad = \langle E_{\mathrm{r}}E_{\mathrm{r}}^*\rangle + \langle E_{\mathrm{l}}E_{\mathrm{l}}^*\rangle \tag{10.30}$$

$$Q = \langle 2A_{\mathrm{r}}A_{\mathrm{l}}\cos\delta_{\mathrm{rl}}\rangle \qquad = \langle E_{\mathrm{r}}E_{\mathrm{l}}^*\rangle + \langle E_{\mathrm{l}}E_{\mathrm{r}}^*\rangle \tag{10.31}$$

$$U = \langle -2A_{\mathrm{r}}A_{\mathrm{l}}\sin\delta_{\mathrm{rl}}\rangle \qquad = \mathrm{i}\left(\langle E_{\mathrm{r}}E_{\mathrm{l}}^*\rangle - \langle E_{\mathrm{l}}E_{\mathrm{r}}^*\rangle\right) \tag{10.32}$$

$$V = \langle A_{\mathrm{r}}^2\rangle - \langle A_{\mathrm{l}}^2\rangle \qquad = \langle E_{\mathrm{r}}E_{\mathrm{r}}^*\rangle - \langle E_{\mathrm{l}}E_{\mathrm{l}}^*\rangle. \tag{10.33}$$

The right hand sides in Eqs. (10.24)–(10.27) and (10.30)–(10.33) only contain terms that are produced by the correlator of a single dish antenna, or equivalently, images of the sky brightness corresponding to the correlator products of an interferometric array.

The Stokes visibilities are the Fourier transforms of the Stokes parameter images of the sky. Because of the linearity of the Fourier transform, they are the sums of the Fourier transforms of the individual terms in the right hand sides of Eqs. (10.24)–(10.27) and (10.30)–(10.33). Remember that an individual point in the uv plane is the cross correlation between two antennas. Dropping the Es and $\langle\cdot\rangle$ for notational clarity, using two distinct antennas 1 and 2, the Stokes visibilities are therefore:

$$\mathscr{I} = x_1x_2^* + y_1y_2^* \qquad = r_1r_2^* + l_1l_2^* \tag{10.34}$$

$$\mathscr{Q} = x_1x_2^* - y_1y_2^* \qquad = r_1l_2^* + l_1r_2^* \tag{10.35}$$

$$\mathscr{U} = x_1 y_2^* + y_1 x_2^* \qquad = \mathrm{i}\left(r_1 l_2^* - l_1 r_2^*\right) \tag{10.36}$$

$$\mathscr{V} = -\mathrm{i}\left(x_1 y_2^* - y_1 x_2^*\right) \qquad = r_1 r_2^* - l_1 l_2^*. \tag{10.37}$$

10.5 Jones Matrices

Until here we have assumed that nothing happens to the polarization state between emission and imaging. In this section we relax this restriction by assuming that all polarization distortions are linear operators, represented by 2×2 complex valued matrices called Jones matrices (Jones 1941). See Hamaker et al. (1996), Sault et al. (1996), Hamaker and Bregman (1996), Hamaker (2000), and Hamaker (2006) for background on their application to radio astronomical data.

Let us first write cross correlation in a convenient way. Because most of the treatment in this section is basis independent, from here on the generic polarizations p and q may designate either x and y, or r and l. As illustrated by the white boxes in Fig. 10.3, radio antenna i's polarizer produces a polarization vector

$$\mathbf{e}_i = \begin{pmatrix} p_i \\ q_i \end{pmatrix}. \tag{10.38}$$

The correlator multiplies antenna i's signal with the complex conjugate of antenna j's. In vector terms, a complex conjugate corresponds to the conjugate transpose (†). The correlator therefore produces a 2×2 complex valued coherency matrix

$$\mathbf{E}_{ij} = \mathbf{e}_i \mathbf{e}_j^\dagger = \begin{pmatrix} p_i \\ q_i \end{pmatrix} \left(p_j^*, q_j^*\right) \tag{10.39}$$

$$\mathbf{E}_{ij} = \begin{pmatrix} p_i p_j^* & p_i q_j^* \\ q_i p_j^* & q_i q_j^* \end{pmatrix}. \tag{10.40}$$

If the polarization distortion between the source and correlator input i can be described by the linear operator (Jones matrix) $\mathbf{J}_i$, then

$$\mathbf{e}_i' = \mathbf{J}_i \mathbf{e}_i, \tag{10.41}$$

where $\mathbf{e}_i$ is the original, and $\mathbf{e}_i'$ is the distorted signal. Cross correlation of the distorted signals yields

$$\mathbf{E}_{ij}' = \mathbf{e}_i' \mathbf{e}_j'^\dagger = \mathbf{J}_i \mathbf{e}_i \left(\mathbf{J}_j \mathbf{e}_j\right)^\dagger = \mathbf{J}_i \mathbf{e}_i \mathbf{e}_j^\dagger \mathbf{J}_j^\dagger = \mathbf{J}_i \mathbf{E}_{ij} \mathbf{J}_j^\dagger. \tag{10.42}$$

Equation (10.42) is called the measurement equation for baseline i–j. Because it is invertible, we can accurately calibrate radio interferometers:

$$\mathbf{E}_{ij} = \mathbf{J}_i^{-1}\mathbf{E}'_{ij}\mathbf{J}_j^{\dagger-1}. \tag{10.43}$$

Note that $\mathbf{J}_i$ may in fact be the product of many effects along the signal path, in right-to-left order of appearance as seen from source to correlator input. Order is important here. Matrix multiplication does not commute. In this following example:

$$\mathbf{J} = \mathbf{GPTFS}, \tag{10.44}$$

the signal first encounters **S** (for example interstellar Faraday rotation), then **F** (for example ionospheric Faraday rotation), **T** (for example atmospheric delays), **P** (for example parallactic angle or antenna element beam), and finally **G** (for example low noise amplifier gain/cross talk and coax cable losses).

Let us look at a few example Jones matrices. A perfect instrument is of course described by the identity matrix

$$\mathbf{J} = \begin{pmatrix} 1 & 0 \\ 0 & 1 \end{pmatrix}. \tag{10.45}$$

A polarization dependent time delay looks like

$$\mathbf{J} = \begin{pmatrix} \mathrm{e}^{2\pi\mathrm{i}\nu\tau_\mathrm{p}} & 0 \\ 0 & \mathrm{e}^{2\pi\mathrm{i}\nu\tau_\mathrm{q}} \end{pmatrix}, \tag{10.46}$$

where τ_p is the delay for the p polarization, and ν is the observing frequency. Clean receiver gain (no leakage between polarizations) is described by the diagonal matrix

$$\mathbf{J} = \begin{pmatrix} g_\mathrm{p} & 0 \\ 0 & g_\mathrm{q} \end{pmatrix}. \tag{10.47}$$

If a certain fraction $d_{\mathrm{q}\to\mathrm{p}}$ of the q signal leaks into p, and a fraction $d_{\mathrm{p}\to\mathrm{q}}$ of the p signal leaks into q, this becomes

$$\mathbf{J} = \begin{pmatrix} g_\mathrm{p} & d_{\mathrm{q}\to\mathrm{p}} \\ d_{\mathrm{p}\to\mathrm{q}} & g_\mathrm{q} \end{pmatrix}. \tag{10.48}$$

Parallactic angle—or any other rotation of the feed system with respect to the IAU definitions of x and y by an angle θ measured north through east—in the Cartesian basis:

$$\mathbf{J} = \begin{pmatrix} \cos\theta & \sin\theta \\ -\sin\theta & \cos\theta \end{pmatrix}. \tag{10.49}$$

In the circular case, a rotation of the feed by an angle θ counter clockwise, corresponds to an rl phase difference of twice the angle θ, see also Fig. 10.1:

$$\mathbf{J} = \begin{pmatrix} e^{-i\theta} & 0 \\ 0 & e^{+i\theta} \end{pmatrix}. \tag{10.50}$$

10.6 Faraday Rotation

When electromagnetic radiation propagates through a magnetized plasma, the left- and right circularly polarized components propagate at different speeds at different wavelengths (Jackson and Jackson 1962). This frequency dependent delay leads to a frequency dependent rotation of the polarization position angle of linearly polarized waves, called Faraday rotation, or the Faraday effect. For a single source along the line of sight with no internal structure, the polarization angle changes proportionally to the square of the wavelength

$$\chi(\lambda) = \chi_0 + \mathrm{RM}\lambda^2, \tag{10.51}$$

where χ is the polarization angle, χ_0 is the polarization angle at infinite frequency (zero wavelength), and λ is the observing wavelength. The parameter $\mathrm{RM} = \frac{\partial \chi}{\partial \lambda^2}$ is the rotation measure of the source (Fig. 10.4).

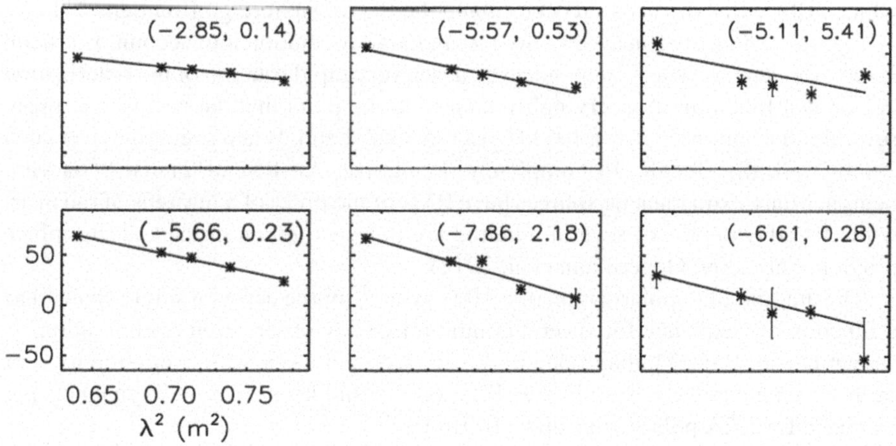

Fig. 10.4 Example of linear dependence of polarization angle on λ^2. Shown are six lines of sight through the polarized Galactic synchrotron foreground, observed with the WSRT at 350 MHz. Credit: Haverkorn et al., A&A, 403, 1031, 2003, reproduced with permission © ESO

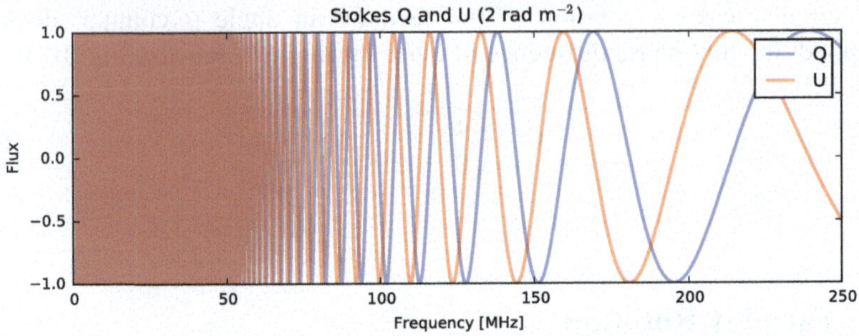

Fig. 10.5 Stokes Q and U for a source at a Faraday depth of 2 rad m^{-2} as a function of observing frequency

More generally, the rotation measure is known as ϕ, or the Faraday depth,

$$\phi = 0.812 \int_{\text{there}}^{\text{here}} n_e \mathbf{B} \cdot \mathbf{dl}, \tag{10.52}$$

which is the electron-density weighted integral of the line-of-sight magnetic field from the source to the observer. Typically, there can be multiple optically thin sources along the line of sight, each at their own Faraday depth. Sources themselves also have a certain Faraday thickness associated with them: the difference in Faraday depth between the front and the back of the source (Burn 1966).

Multi-frequency polarimetry provides insight into the plasma properties of the source and the interstellar medium. In rare cases one may even derive the order in which diffuse bits of emission occur along the line of sight (e.g., Brentjens 2011).

Figure 10.5 nicely illustrates why taking Faraday rotation into account is crucial at low frequencies. The λ^2 dependence causes very rapid rotation of the polarization vector as a function of observing frequency, to the point that the vector may make several full rotations across a 195 kHz LOFAR sub band. When averaging over such a sub band, the signal will completely depolarize. An RM of 2 rad m^{-2} is very benign. Many extra galactic sources have RMs of the order of a few tens of rad m^{-2}, while Faraday depths of several hundred are not uncommon in the Galactic plane (Taylor et al. 2009; Oppermann et al. 2012).

The amount of depolarization of a 100 rad m^{-2} source across a single channel as a function of frequency for several common LOFAR observation configurations is shown in Fig. 10.6. Although one may very well get away with averaging to 4, or even 1 channel per sub band in the HBA, one should keep *at least* 16 channels per sub band for LBA polarimetry up to 100 rad m^{-2}.

In most cases, however, detecting faint sources requires combining tens of MHz of bandwidth. To prevent depolarization, one can record the data in narrow channels and combine the channels in a clever way later. If one constructs a complex polarization $P = Q + \mathrm{i}U$ (Fig. 10.7), and the frequency channels are sufficiently narrow, there is a (conditionally) invertible Fourier relation between the complex polarization as a function of λ^2 and the linearly polarized emissivity as a function

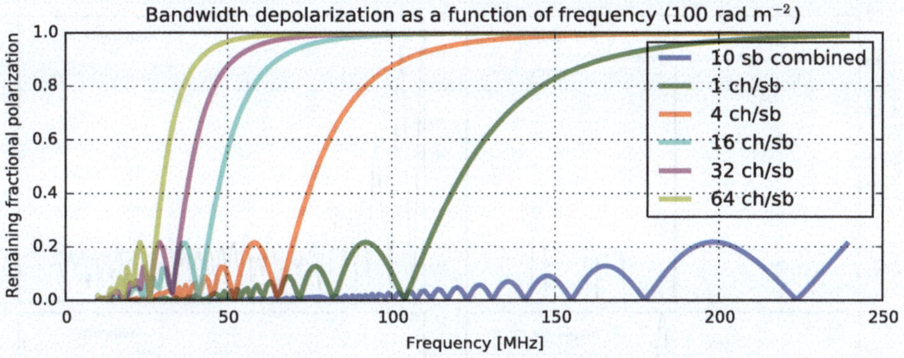

Fig. 10.6 Bandwidth depolarization of a 100 rad m^{-2} source as a function of observing frequency. The plot shows curves for various common LOFAR frequency averaging settings. One sub band (the green curve) is assumed to be 195.3125 kHz wide

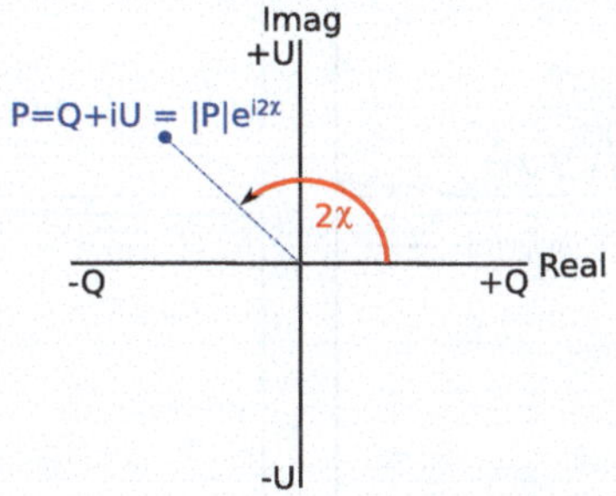

Fig. 10.7 Complex polarization $Q + \mathrm{i}U$. χ is the north-through-east polarization angle on the sky

of Faraday depth ϕ (Burn 1966; Brentjens and de Bruyn 2005):

$$P(\lambda_c^2) = \int_{\lambda^{2'}} \int_{\Omega} \int_{-\infty}^{\infty} W(\Omega, \lambda_c^2 + \lambda^{2'}) f(\Omega, \phi, \lambda_c^2 + \lambda^{2'}) \mathrm{e}^{2\mathrm{i}\phi(\lambda_c^2+\lambda^{2'})}\, \mathrm{d}\phi\, \mathrm{d}\Omega\, \mathrm{d}\lambda^{2'} \quad (10.53)$$

$$P(\lambda^2) \approx W(\lambda^2) \int_{-\infty}^{\infty} f(\phi) \mathrm{e}^{2\mathrm{i}\phi\lambda^2}\, \mathrm{d}\phi \quad (10.54)$$

$$\tilde{f}(\phi) = f(\phi) * R(\phi) \approx \int_{-\infty}^{\infty} P(\lambda^2) \mathrm{e}^{-2\mathrm{i}\phi\lambda^2}\, \mathrm{d}\lambda^2. \quad (10.55)$$

The above relations are illustrated in Fig. 10.8. In Eqs. (10.53)–(10.55), Ω represents solid angle, $f(\phi)$ is known variously as the Faraday dispersion function, Faraday spectrum, or RM spectrum, $*$ denotes convolution and $R(\phi)$ is called the RM spread function (RMSF), analogous to the point spread function (PSF) in imaging. Schnitzeler and Lee (2015) rewrite Eq. (10.53) as an integral over frequency instead of λ^2, which allows them to relax the assumption of narrow channels, and leads to a slightly modified version of Eq. (10.54) that is more accurate

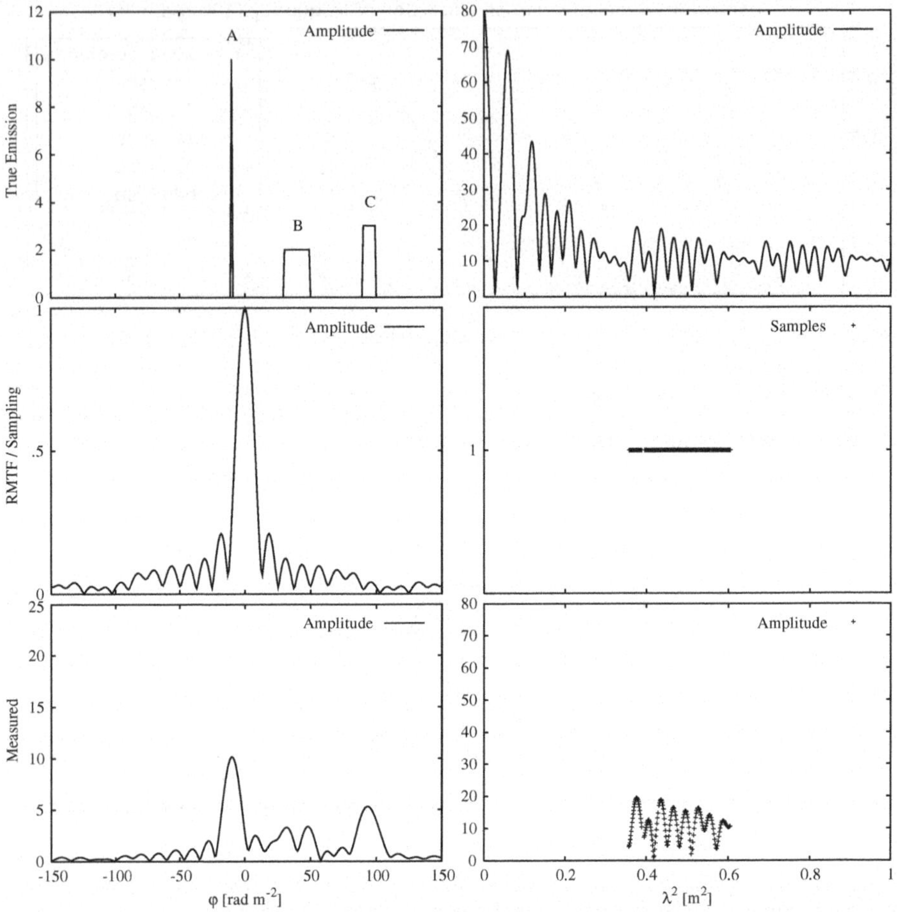

Fig. 10.8 An example of multiple radio-emitting, and Faraday rotating sources along the line of sight. The left column is in Faraday space, the right column in λ^2 space. A Fourier transform relates left and right per Eq. (10.54). The bottom panels are the convolution of the top two (left), or the product of the top two (right). Credit: Brentjens and de Bruyn, A&A, 441, 1217, 2005, reproduced with permission © ESO

when the finite channel width can not be ignored anymore for certain Faraday depths, such as in LBA observations.

Because one knows the RMSF, one can attempt to deconvolve $\tilde{f}(\phi)$ (Heald et al. 2009). Although possible, this is less straightforward than it naively appears; the "imaging" domain (ϕ space) is complex valued, causing "RM clean" components to interfere with each other in complicated ways. Sun et al. (2015) analyse this in detail and compare various algorithms for proper $f(\phi)$ reconstruction.

Because all pixels in a Q and U image pair have been observed at the same frequency, every line of sight in a cube of $Q + iU$ channel maps uses the *same* Fourier transform to compute $\tilde{f}(\phi)$. The process is therefore very fast. The result of doing this transformation on a cube of channel maps is a cube of maps of the polarized emission as a function of Faraday depth: a rotation measure cube, or RM-cube. Because each frame is the phase-weighted mean of all channel maps, each frame has a noise level corresponding to that obtained when averaging the full bandwidth. A Python program to compute a "dirty" RM-cube ($\tilde{f}(\phi, \alpha, \delta)$) from Q and U FITS images is available for download at https://github.com/brentjens/rm-synthesis.

10.7 Ionosphere

As introduced in Chap. 7, the interstellar medium (ISM) is not the only magnetized plasma encountered by cosmic radio waves. Although the path length through the Earth's ionosphere is tiny compared to the many kpc of ISM that are typically traversed, its electron density and magnetic field are many orders of magnitude higher. Figure 10.9 shows a map of the total ionospheric electron content (TEC) derived from GPS satellite data. Such maps (images as well as in machine-readable form) are published by various organizations, such as the Royal Observatory of Belgium,[2] and the Centre for Orbit Determination in Europe[3] (CODE) of the university of Bern. The right panel shows the observed variations in the RM towards a pulsar, plotted on top of an RM derived by combining TEC maps with a global geomagnetic field model. They match nicely. Note the gradual decline during the night when the ionosphere recombines, and the sudden jump around sunrise when there are plenty of UV photons again: the ionosphere's daily epoch of reionization.

The variations seen in Fig. 10.9 are fairly typical. In the Netherlands we see daily variations anywhere from ± 0.2 to $\pm 5\,\mathrm{rad\,m^{-2}}$. In other places around the world, variations up to $\pm 10\,\mathrm{rad\,m^{-2}}$ are possible. If not corrected, these variations cause depolarization when combining data from an entire day or night. A $1\,\mathrm{rad\,m^{-2}}$ change rotates the plane of polarization by 4 rad at 150 MHz (2 m wavelength), leading to full depolarization, and things get worse rapidly towards lower frequencies.

There are various approaches to correct LOFAR data for these variations. The easiest is a bright, highly polarized calibrator inside the field of view within a few degrees of the target. One determines the ionospheric RM variations of the calibrator, corrects the visibilities using a Jones matrix very similar to Eq. (10.49),[4] and images again. The accuracy is entirely determined by the amount of linearly polarized flux in the calibrator. Sadly, at LOFAR frequencies only a handful of

[2] http://gnss.be/Atmospheric_Maps/ionospheric_maps.php.

[3] http://aiuws.unibe.ch/ionosphere/.

[4] What form does it have, exactly? Derive it yourself, but be careful about directions and signs!

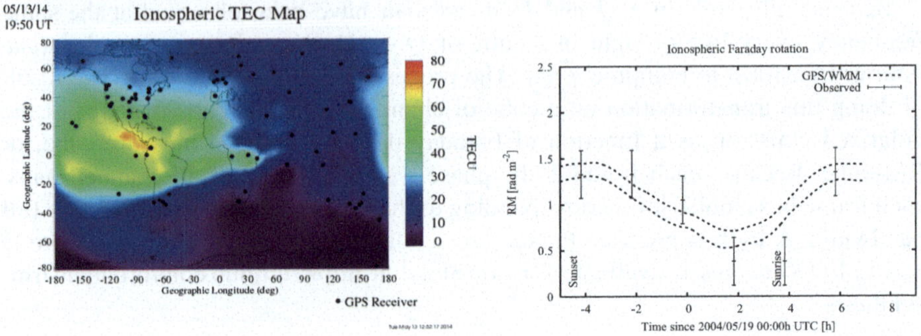

Fig. 10.9 Left: total electron content of the ionosphere as derived from GPS satellite signal delays (http://iono.jpl.nasa.gov/latest_rti_global.html). Right: WSRT observation of variations in RM towards a pulsar, compared to expected ionospheric RMs based on GPS-derived ionospheric electron contents and a model of the geomagnetic field (world magnetic model, WMM). Right hand figure credit: Brentjens, A&A, 489, 69, 2008, reproduced with permission © ESO

pulsars and even fewer radio galaxies are bright enough to deal with typical ionospheres (e.g. Mulcahy et al. 2014).

As can be seen in Fig. 10.9, the combination of GPS TEC data with geomagnetic field models matches the data very well. Typical accuracies are of the order of $0.2\,\mathrm{rad\,m^{-2}}$, which is good enough to prevent depolarization due to residuals in the HBA, but is insufficient for the LBA. More information about these methods can be found in Sotomayor-Beltran et al. (2013), Brentjens (2008), and the documentation of the AIPS command FARAD.

A variation on the first scheme is to use the diffuse Galactic polarized synchrotron radiation as an RM-calibrator, alternately determining the RM-offsets as a function of time, and the actual structure of the Galactic synchrotron emission itself until both converge, somewhat akin to how self calibration is used to alternately solve for telescope parameters and image models of the sky. Accuracies of the order of $10\,\mathrm{mrad\,m^{-2}}$ have been achieved with LOFAR (Brentjens et al. *in prep.*), but the success of this method depends entirely on the availability of diffuse polarized synchrotron radiation in the field of view. Fortunately, we have not yet found fields where it is completely absent. Although the Galactic polarized synchrotron emission is mainly visible at short baselines, the corrections derived from it also apply to the farthest stations once differential Faraday rotation from station to station has been corrected for. As is shown below, the latter is simple.

Figure 10.10 compares an uncorrected HBA RM-cube to the same cube corrected using GPS-TEC models, and corrected through RM-selfcal. Even for the benign ionosphere in this observation, correcting for time variable Faraday rotation increased the brightness of the polarized synchrotron, while it decreased the brightness of the instrumentally polarized sources. The latter occurs because celestial sources are smeared out in Faraday space due to changes in ionospheric RM, while instrumentally polarized sources are not: the instrumental polarization happens

Fig. 10.10 Correcting for ionospheric Faraday rotation. From left to right: no correction, GPS-TEC + geomagnetic field model, RM-selfcal. Colour indicates RM value, and brightness. This observation suffered from only about 0.3 rad m^{-2} changes in ionospheric Faraday rotation, so it was very benign. Data from Jelić et al. (2015)

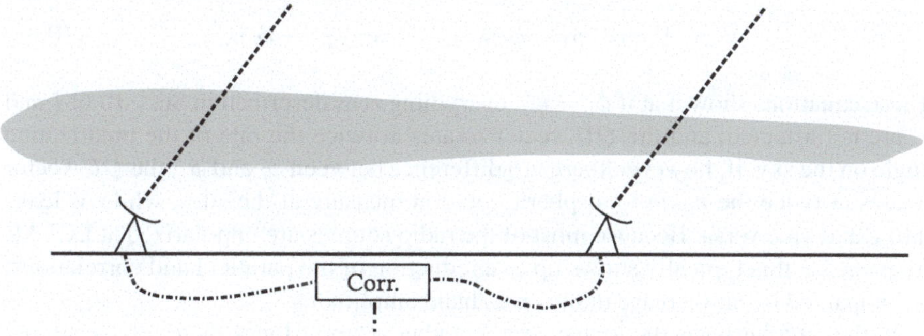

Fig. 10.11 Differential Faraday rotation. Different antenna stations look through very different parts of the ionosphere, each part with its own Faraday rotation. When combining those signals in the cross correlator, parallel hand signals are rotated to the cross hand correlations and vice versa

at the antennas, which are well below the ionosphere. If one now "focuses" the celestial sources in RM-space by applying the time variable ionospheric corrections, one at the same time smears out or "defocuses" the instrumentally polarized sources, reducing their brightness. Ironically, worse ionospheric nights decrease the amount of false, instrumental polarization in the maps!

For interferometers, this is not all there is to ionospheric Faraday rotation. Because antenna stations may be tens or even hundreds to thousands of km apart, they see different parts of the ionosphere, with different amounts of ionospheric Faraday rotation (Fig. 10.11). When signals from those stations are correlated, something interesting happens.

The Jones matrix that describes Faraday rotation in the IAU's Cartesian xy system is

$$\mathbf{J} = \begin{pmatrix} \cos(\phi\lambda^2) & -\sin(\phi\lambda^2) \\ \sin(\phi\lambda^2) & \cos(\phi\lambda^2) \end{pmatrix}. \tag{10.56}$$

The measurement equation for the baseline between antennas i and j then becomes

$$\mathbf{E}' = \begin{pmatrix} \cos(\phi_i\lambda^2) & -\sin(\phi_i\lambda^2) \\ \sin(\phi_i\lambda^2) & \cos(\phi_i\lambda^2) \end{pmatrix} \frac{1}{2} \begin{pmatrix} I+Q & U+\mathrm{i}V \\ U-\mathrm{i}V & I-Q \end{pmatrix} \begin{pmatrix} \cos(\phi_j\lambda^2) & -\sin(\phi_j\lambda^2) \\ \sin(\phi_j\lambda^2) & \cos(\phi_j\lambda^2) \end{pmatrix}^\dagger, \tag{10.57}$$

where $\mathbf{E}'$ is the observed coherency matrix.

Writing this out, and calculating the observed I', Q', U', and V' from $\mathbf{E}'$, we obtain

$$I' = I\cos((\phi_j - \phi_i)\lambda^2) - \mathrm{i}V\sin((\phi_j - \phi_i)\lambda^2) \tag{10.58}$$

$$Q' = Q\cos((\phi_j + \phi_i)\lambda^2) - U\sin((\phi_j + \phi_i)\lambda^2) \tag{10.59}$$

$$U' = U\cos((\phi_j + \phi_i)\lambda^2) + Q\sin((\phi_j + \phi_i)\lambda^2) \tag{10.60}$$

$$V' = V\cos((\phi_j - \phi_i)\lambda^2) - \mathrm{i}I\sin((\phi_j - \phi_i)\lambda^2). \tag{10.61}$$

These equations show that if $\phi_i = \phi_j$, everything is as described in Sect. 10.6: I and V are not affected, and the QU vector rotates at twice the rate of the polarization angle on the sky. If, however, there is a difference between ϕ_i and ϕ_j, the QU vector rotates at twice the *mean* ionospheric rotation measure at the sites, while V leaks into I and vice versa. Because most of the radio sources are unpolarized at LOFAR frequencies, this typically shows up as a reduction of the parallel hand correlations, accompanied by an increase in the cross hand amplitudes.

In the circular basis the ionospheric Faraday rotation Jones matrices are simply diagonal matrices that are for all intents and purposes identical to the electronic gain Jones matrices of circular feeds. That is why VLBI observers usually process their data in the circular basis: a simple electronic gain solution automatically absorbs the differential Faraday rotation.

For LOFAR, differential Faraday rotation becomes important at baselines of only a few tens of km in the LBA and at the longer Dutch—and certainly all international baselines—in the HBA.

10.8 Antenna Beam Polarization

In aperture array telescopes such as LOFAR, the individual feeds are stuck to the ground and do not track the sky. Projection of the dipoles when viewed from anywhere else than the zenith causes significant variations in the dipole antenna's polarization response. The polarization leakage to any target that is not at the zenith is typically of the order of tens of percent. Fortunately, the scale of the polarization leakage beams is similar to that of the total intensity beam of a single antenna element: pretty much the entire sky. A well calibrated station does not introduce any other polarimetric distortions in the analogue and digital beams. Because the

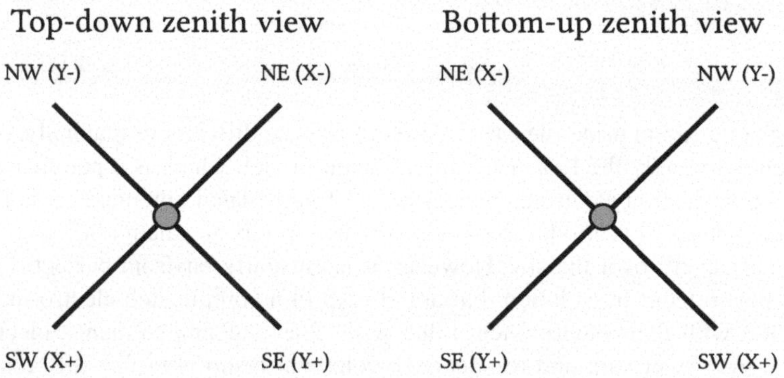

Fig. 10.12 LOFAR antenna polarizations

digital beam is small compared to the element beam, the polarization response does not vary significantly across a station beam. Calibrating only the polarization response towards the pointing centre usually suffices to obtain low leakage wide field maps. This is different from traditional dish antennas, where the on-axis leakage is typically low, but the off-axis leakage rapidly increases towards the edge of the beam.

Figure 10.12 shows two drawings of the polarizations and polarities of LOFAR dipoles. It is immediately clear that they do not adhere to the IAU conventions in Fig. 10.2—not even at the zenith.

The first thing that stands out is the 45° rotation in the station's horizontal plane. In fact, it is not exactly 45° for most stations. The antenna elements are oriented to the so-called reference direction, which bisects the northern angle between the x and y dipoles ("up" in the drawings). It is important to understand that the reference direction only corresponds to geographic north at the centre of CS002LBA. For all other antenna fields, the reference direction is the intersection of the field's ground plane and a plane that is parallel to CS002LBA's meridian plane. This ensures that the dipoles are as parallel as possible, although stations may be far apart and have different field slopes. For fields that have parallel ground planes, the reference directions—and therefore the dipole orientations—are the same. The second thing is much less obvious, but equally important. It is the polarity: the handedness of the LOFAR dipoles is the opposite of the handedness of the IAU convention.

To develop some intuition for working with raw LOFAR data, let us for simplicity's sake look at the IAU Stokes parameters for a source at the zenith of CS002LBA, and compare those to the naive Stokes parameters obtained by blindly applying Eqs. (10.35)–(10.37) to the raw visibilities. The consequences are that

$$I_{\mathrm{raw}} \sim I_{\mathrm{true}} \tag{10.62}$$

$$Q_{\mathrm{raw}} \sim U_{\mathrm{true}} \tag{10.63}$$

$$U_{\mathrm{raw}} \sim -Q_{\mathrm{true}} \tag{10.64}$$

$$V_{\mathrm{raw}} \sim -V_{\mathrm{true}} \tag{10.65}$$

$$\phi_{\mathrm{raw}} = -\phi_{\mathrm{true}}. \tag{10.66}$$

Although one can in principle easily correct for these differences manually, a better approach is to apply the LOFAR element beam model, which is a parametrization of a full-polarization electromagnetic model of an isolated antenna element above a ground plane. This model does a much better job at calculating appropriate Jones matrices all over the sky. However, it is currently far from perfect. LOFAR dipoles never occur in isolation, but are always in a complicated electro-magnetic interaction with their environment filled with other antenna elements, electronics cabinets, trees, wet soil, and roe deer. The element beam is in fact different from antenna to antenna and varies between antennas at the edge of a station and antennas at its centre. Even given this sub-optimal, simplified beam model that assumes that all dipoles are identical, polarization leakage calibration is still rather straightforward because most compact celestial sources are not or only weakly polarized below 300 MHz.

Figure 10.13 shows the response of this element beam model to an unpolarized source. Note that in this simulation, the azimuth was measured with respect to

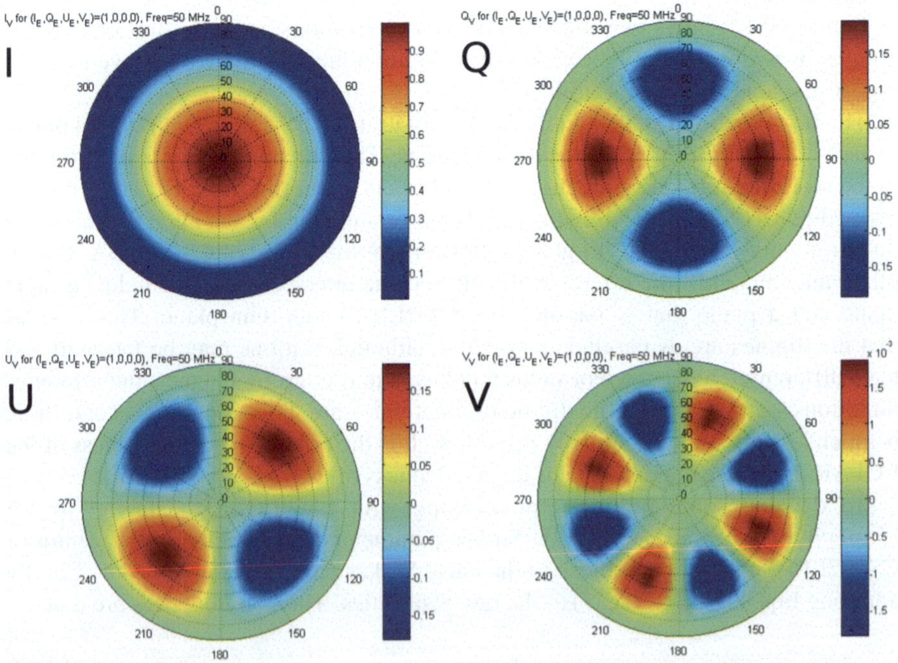

Fig. 10.13 LBA element beam model's response to an unpolarized source. The outer circle is the horizon. The *x* axis is aligned left-to-right, and the *y* axis bottom-to-top: these plots look *down* onto the antenna. Note the differences in colour scale for each Stokes parameter

the x dipole, not the reference direction. Furthermore, it did not compute proper IAU Stokes parameters in the RA/Dec system, but simply filled in Eqs. (10.35)–(10.37) at every azimuth and zenith angle. Fortunately, the element beam model in LOFAR's BBS, (N)DPPP, and AWImager tools takes all these differences into account properly.

Although a well calibrated station should not introduce any additional polarization distortions into the digital beam shape, these distortions do occur in practice. The Jones matrix of a digital beam in a certain direction is the sum of the Jones matrices of all the individual elements, where each element's beam model is multiplied by a certain position-independent calibration factor that takes into account gain differences and cable delays:

$$\mathbf{J} = \left(\sum_{i=1}^{N} \mathbf{G}_i \right) \mathbf{P}(\mathrm{az}, \mathrm{el}), \tag{10.67}$$

where $\mathbf{P}(\mathrm{az}, \mathrm{el})$ is the element beam pattern, and $\mathbf{G}_i$ contains the differences from a perfect calibration. A phase difference between x and y dipoles, or a mis-calibrated electronic gain leads to a polarization distortion, which is easy to determine using standard LOFAR self calibration by solving for all elements in the Jones matrix. These mis-calibration problems are very stable on time scales of weeks to months. For tied array beam forming this is more of a problem because the Jones matrices of different stations may be so different that the signal to noise ratio of the coherent sum (tied array) of the core stations is degraded. The remaining polarimetric distortion should nevertheless still be calibratable by observing pulsars with known polarimetric properties.

10.9 Summary

Radio antennas are fundamentally polarized detectors, and polarimetry is required to observe certain astrophysical objects and processes. For aperture arrays like LOFAR, a good understanding of polarimetry is even required to make high quality images of unpolarized sources. Fortunately, most of the math is straightforward, and in many cases already implemented in dedicated LOFAR calibration and imaging tools.

Although LOFAR's polarization leakage towards most targets is rather prominent by parabolic dish standards, it hardly changes across the field of view of an HBA station, making wide field polarimetry with the HBA surprisingly easy. Because the LBA's digital beams are larger than the HBA's, there is some clear variation across the LBA field of view.

The biggest enemy of low frequency polarimetry is ionospheric Faraday rotation. There are several methods available to determine the required corrections either from the data itself, or by using external models. All these methods work well in the HBA. What works for the LBA is a matter of ongoing research.

Finally, a word of caution. In polarimetry, signs and conjugations matter a lot. Always be vigilant when reading polarimetry texts, and never assume the authors used one or the other sign convention, even if they said so. Sign errors are easy to make and hard to find. Always verify everything yourself. Particularly in this chapter.

References

Born, M., Wolf, E.: Principles of Optics, 7th edn. Cambridge University Press, Cambridge (1999)
Bracewell, R.: The Fourier Transform and Its Applications. McGraw-Hill International Editions. McGraw Hill, New York (2000)
Brentjens, M.A.: Astron. Astrophys. **489**, 69 (2008)
Brentjens, M.A.: Astron. Astrophys. **526**, A9 (2011)
Brentjens, M.A., de Bruyn, A.G.: Astron. Astrophys. **441**, 1217 (2005)
Burn, B.: Mon. Not. R. Astron. Soc. **133**, 67 (1966)
Chandrasekhar, S.: Astrophys. J. **104**, 110 (1946)
Cotton, W.D.: In: Taylor, G.B., Carilli, C.L., Perley, R.A. (eds.) Synthesis Imaging in Radio Astronomy II. Astronomical Society of the Pacific Conference Series, vol. 180, p. 111 (1999)
Hamaker, J.P.: Astron. Astrophys. Suppl. Ser. **143**, 515 (2000)
Hamaker, J.P.: Astron. Astrophys. **456**, 395 (2006)
Hamaker, J.P., Bregman, J.D.: Astron. Astrophys. Suppl. Ser. **117**, 161 (1996)
Hamaker, J.P., Bregman, J.D., Sault, R.J.: Astron. Astrophys. Suppl. Ser. **117**, 137 (1996)
Haverkorn, M., Katgert, P., de Bruyn, A.G.: Astron. Astrophys. **403**, 1031 (2003)
Heald, G., Braun, R., Edmonds, R.: Astron. Astrophys. **503**, 409 (2009)
IAU: Transactions of the IAU, vol. 158, p. 166 (1973)
Jackson, J.D., Jackson, J.D.: Classical Electrodynamics, vol. 3. Wiley, New York (1962)
Jelić, V., de Bruyn, A.G., Pandey, V.N., et al.: Astron. Astrophys. **583**, A137 (2015)
Jones, R.C.: J. Opt. Soc. Am. **31**, 488 (1941)
Mulcahy, D.D., Horneffer, A., Beck, R., et al.: Astron. Astrophys. **568**, A74 (2014)
Oppermann, N., Junklewitz, H., Robbers, G., et al.: Astron. Astrophys. **542**, A93 (2012)
Sault, R.J., Hamaker, J.P., Bregman, J.D.: Astron. Astrophys. Suppl. Ser. **117**, 149 (1996)
Schnitzeler, D.H.F.M., Lee, K.J.: Mon. Not. R. Astron. Soc. **447**, L26 (2015)
Sotomayor-Beltran, C., Sobey, C., Hessels, J.W.T., et al.: Astron. Astrophys. **552**, A58 (2013)
Stokes, G.G.: Trans. Camb. Philos. Soc. **9**, 399 (1852)
Sun, X.H., Rudnick, L., Akahori, T., et al.: Astron. J. **149**, 60 (2015)
Taylor, G.B., Carilli, C.L., Perley, R.A., (eds.): Synthesis Imaging in Radio Astronomy II. Astronomical Society of the Pacific Conference Series, vol. 180 (1999)
Taylor, A.R., Stil, J.M., Sunstrum, C.: Astrophys. J. **702**, 1230 (2009)
Thompson, A.R., Moran, J.M., Swenson, G.W.: Interferometry and Synthesis in Radio Astronomy, 2nd edn. Wiley-VCH, Weinheim (2001)
van Haarlem, M.P., Wise, M.W., Gunst, A.W., et al.: Astron. Astrophys. **556**, A2 (2013)

Chapter 11
Long Baseline Imaging with LOFAR

Javier Moldón and Eskil Varenius

Abstract In this chapter we focus on the calibration of International LOFAR, which includes the long baselines provided by international stations, to produce high-resolution radio images. The Very Long Baseline Interferometry (VLBI) techniques are explained, as well as the different steps required to properly calibrate a long-baseline observation at low frequencies.

11.1 Introduction

The prime reason to include the international LOFAR stations in the array is to obtain very high-resolution images. Using the longest LOFAR baselines, subarcsecond imaging is possible with the High Band Antenna (HBA) and the upper frequency range of the Low Band Antenna (LBA). Early science results include images of AGN jets, and of individual supernova remnants in M82 (see Fig. 11.1). In this chapter we describe the international LOFAR stations and, in general terms, some important things to keep in mind when observing, calibrating and imaging data including *long* baselines, i.e. baselines to international stations. In particular we summarise how techniques from Very Long Baseline Interferometry (VLBI) can be used to calibrate LOFAR data using the longest baselines. Some parts of the text are focused on HBA data; calibration and imaging of long baseline data from

J. Moldón (✉)
Jodrell Bank Centre for Astrophysics, School of Physics and Astronomy, The University of Manchester, Manchester, UK

ASTRON, Dwingeloo, The Netherlands
e-mail: javier.moldon@manchester.ac.uk

E. Varenius
Onsala Space Observatory, Department of Earth and Space Sciences, Chalmers University of Technology, Onsala, Sweden
e-mail: varenius@chalmers.se

G. Heald et al. (eds.), *Low Frequency Radio Astronomy and the LOFAR Observatory*, Astrophysics and Space Science Library 426,
https://doi.org/10.1007/978-3-319-23434-2_11

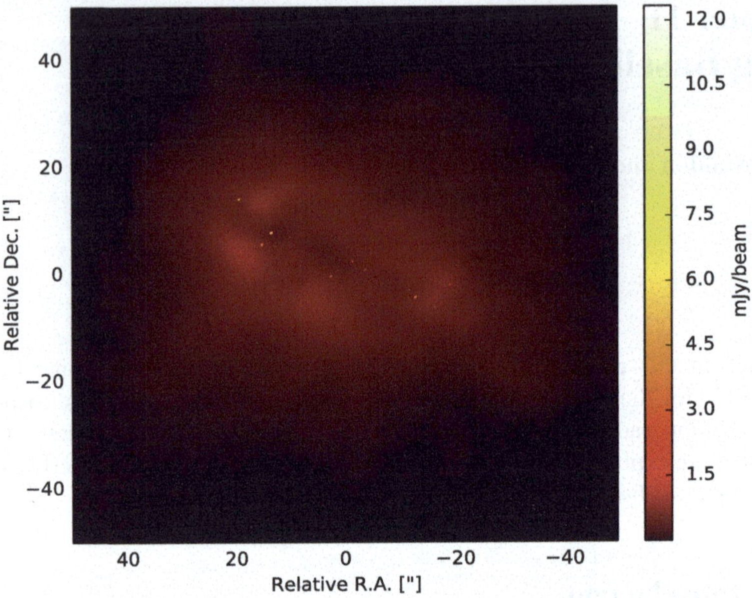

Fig. 11.1 Combined high resolution and low resolution image of M82 illustrating the relative brightness between the compact and extended emission at 154 MHz. The synthesized beam size of the International LOFAR image is 0.36″ × 0.23″ and has an rms noise level of 0.15 mJy/beam. Credit: Varenius et al., A&A, 574, 114, 2015, reproduced with permission © ESO

the LBA is more challenging and additional work is still needed to find the best approach.

The outline of this chapter is as follows: in Sect. 11.2 we describe the international LOFAR stations and what to expect in general in terms of resolution, sensitivity and field of view. In Sect. 11.3 we describe how to calibrate visibility phases and amplitudes using data on the longest LOFAR baselines, in particular how to deal with residual delays and rates. In Sect. 11.4 we summarise the optimal observing strategy for enabling calibration of the international stations. Finally in Sect. 11.5 we discuss a few practical considerations which may be useful when working with LOFAR data including international stations.

11.2 The International LOFAR Stations

The majority of the LOFAR stations, namely the core and remote stations, are distributed over an area roughly 180 km in diameter predominantly in the northeastern Dutch province of Drenthe. Currently, the array also includes 13 international LOFAR stations across Europe that provide maximum baselines up to 1980 km. A further station is planned to be built in Latvia. Figure 11.2 shows the distribution of stations that already exist.

Fig. 11.2 LOFAR is composed of 24 core stations and 14 remote stations in the Netherlands, and 13 international stations (+1 planned)

11.2.1 Sampling of Fourier Space

The core stations provide maximum baselines of 2.7 km, the remote stations of 120 km, and the international stations of 1980 km. Table 11.1 shows the distances between each pair of international stations, including the core station CS001 as a reference to the centre of the array. Since there are relatively few international stations, the sampling of Fourier space (uv-coverage) is less dense for baselines to international stations compared to core- and remote baselines. A typical LOFAR uv coverage is shown in Fig. 11.3. It should be noted that because of the wide bandwidth offered by LOFAR, Multi-Frequency-Synthesis techniques can be used in imaging to provide very good uv-coverage also at the longest baselines.

11.2.2 Resolution and Sensitivity

In general, the image resolution from interferometric data depends on the sampling of Fourier space and relative weighting applied to the visibilities when imaging. A quick estimate can however be obtained using the well known expression $\theta \approx \lambda/D$ where λ is the wavelength of the observation, D is the maximum baseline length, and θ is the angular resolution in radians. For international LOFAR baselines we estimate an angular resolution of about $0.2''$ at 150 MHz. This is also verified in

Table 11.1 Separation in km between the Dutch core (represented by the station CS001) and the currently available international stations

	CS001	DE609	DE605	DE601	DE603	DE604	DE602	SE607	UK608	FR606	IE613	PL610	PL611	PL612
CS001	0	225	226	266	396	419	581	594	602	700	990	694	999	917
DE609	225	0	394	412	325	248	585	430	825	892	1187	502	834	701
DE605	226	394	0	53	372	487	440	807	552	498	1012	753	1002	1010
DE601	266	412	53	0	344	476	390	833	590	490	1057	734	970	997
DE603	396	325	372	344	0	186	277	714	920	800	1361	398	633	671
DE604	419	248	487	476	186	0	455	556	1005	957	1407	277	590	524
DE602	581	585	440	390	277	455	0	990	959	690	1439	587	689	862
SE607	594	430	807	833	714	556	990	0	1110	1292	1342	658	1000	691
UK608	602	825	552	590	920	1005	959	1110	0	495	494	1280	1550	1515
FR606	700	892	498	490	800	957	690	1292	495	0	960	1197	1372	1468
IE613	990	1187	1012	1057	1361	1407	1439	1342	494	960	0	1679	1981	1880
PL610	694	502	753	734	398	277	587	658	1280	1197	1679	0	351	278
PL611	999	834	1002	970	633	590	689	1000	1550	1372	1981	351	0	404
PL612	917	701	1010	997	671	524	862	691	1515	1468	1880	278	404	0

Stations in Germany have names beginning with DE, those in Sweden with SE, the UK station starts with UK, the French station starts with FR, the Irish station with IE, and the Polish stations start with PL

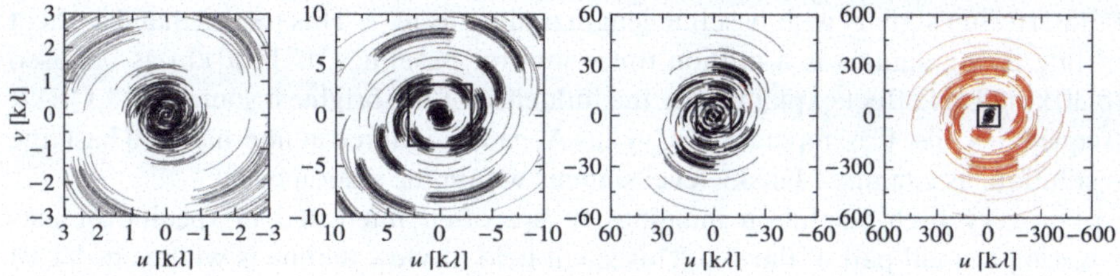

Fig. 11.3 *uv* coverage for a 4-h observation of a source at declination +48° with a single subband centred at 140 MHz. Only one visibility every 160 s is shown. The rectangles in the last three panels show the area covered by the previous panel. Visibilities corresponding to baselines with international stations are plotted in red

practice, for example in the observations of M82 (Varenius et al. 2015). Estimates for LBA and HBA can be found in Sect. 11.2.3.1.

The expected image noise depends on many factors, such as which baselines to include in the final imaging, but also on the solar activity at the time of observation. For subarcsecond imaging it is common to exclude the NL-baselines and only include data on baselines to international stations. LOFAR HBA is most sensitive at frequencies around 150 MHz (van Haarlem et al. 2013, Fig. 22). Varenius et al. (2015) obtained rms noise levels of 0.3 mJy/beam at 118 MHz and 0.15 mJy/beam at 154 MHz using 16 MHz bandwidth and 16 h of integration, in reasonable agreement with theoretical estimates (see Varenius et al. (2015) for a brief discussion on the expected thermal noise). Since then, more international stations have became available and station calibration has been improved. It is reasonable to assume similar rms noise levels (scaling with bandwidth and integration time) can be expected in future observations using international baselines.

11.2.3 Field of View

The sky area possible to image from any LOFAR observation is limited by factors such as the station beam, baseline projection effects, and atmospheric disturbances across the sky. NL-LOFAR observations are often also limited by very bright interfering sources, such as the A-team. International baseline imaging is in many aspects simpler than Dutch baseline imaging, because we can generally ignore interference from other bright sources in the sky.

For a distant potentially interfering source, the interfering contribution to the target source visibility declines by factors of u^{-1} as a function of baseline length (u) for both frequency and time decorrelation (also called *smearing*; see Chap. 6 and Sect. 11.2.3.1). Furthermore, for most of the source population resolved on longer Dutch LOFAR baselines, the intrinsic visibility structure decreases faster than u^{-2} (a rough approximation based on experience of typical dependence of visibility amplitude vs baseline length for resolved radio sources), giving a total decrease of interfering signals with baseline length scaling as u^{-4}. This means that the effect of interfering signals is a million times less for baselines of 1000 km as opposed to 30 km. This effect explains why the influence of the brightest sources at LOFAR frequencies, like Cassiopeia A or Cygnus A, can be ignored at international baseline resolution, as can most Jansky level sources within the station beam.

For very high-resolution imaging we are often interested in specific objects covering a small part of the sky. This small field of view regime is where cm-VLBI usually operates and this regime greatly simplifies imaging. In this regime, target images are generally smaller than the isoplanatic patch, and therefore only a single-direction station-dependent correction needs to be determined. Likewise, over such small fields, w-term effects and station beam variations can generally be ignored. This is different from LOFAR core-resolution imaging, where in order to get noise-limited (rather than "dynamic range"-limited) images at any given point in a field the whole field must be imaged using multi-directional calibration techniques.

Whether one can produce useful images using the small-field approximation and VLBI software depends on the brightness of the target source and, if necessary, the availability of close sources to use as calibrators similar to what is done in cm-VLBI.

11.2.3.1 Time and Frequency Smearing

Although smearing helps to simplify calibration by reducing the influence of bright interfering sources, care has to be taken to not average too much so that the science targets are affected. In this section we estimate the impact of smearing on the field of view.

For LOFAR, the standard *raw* data are delivered from the correlator with resolution 1 s in time, and 64 channels per subband. Each subband (using the standard 200 MHz clock) is 195 kHz wide, meaning that the default minimum averaging bandwidth is 3 kHz. This will limit the dynamic range at some distance from the observed phase centre, similar to the limit imposed by the station beam. A detailed description of the averaging losses is beyond the scope of this chapter, we merely quote the often-used results by Taylor et al. (1999, chapter 18), who derived two expressions to estimate the average amplitude loss due to averaging in frequency and time, at some distance from the phase centre. For frequency smearing, we can use their expression 18–24 (see also Chap. 6 of this volume) assuming a square bandpass and circular Gaussian tapering, where the reduction in amplitude at a distance from the phase centre r can be estimated as

$$\frac{I}{I_0} = \frac{\sqrt{\pi}}{2\sqrt{\ln 2}}\frac{\theta \nu_c}{r \Delta\nu}\,\mathrm{erf}\left(\sqrt{\ln 2}\frac{r\Delta\nu}{\theta\nu_c}\right) \tag{11.1}$$

where θ is the synthesized beam size (FWHM), ν_c is the central frequency of the observation, and $\Delta\nu$ is the bandwidth. Note that the units of θ and r cancel if they are given in the same unit. Note also that this expression is in fact independent of

Table 11.2 Station FWHM Values taken from van Haarlem et al. (2013, App. B)

Freq. (MHz)	λ (m)	Int. PSF FWHM (″)	Int. station FWHM (deg)	5% loss, 1s Diam. (deg)	5% loss, 64ch/SB Diam. (deg)	5% loss, 2s Diam. (deg)	5% loss, 4ch/SB Diam. (deg)
15	19.99	1.67	19.39	5.92	2.16	2.96	0.14
30	9.99	0.83	9.70	2.96	2.16	1.48	0.14
45	6.66	0.56	6.46	1.98	2.16	0.98	0.14
60	5.00	0.42	4.85	1.48	2.16	0.74	0.14
75	4.00	0.33	3.88	1.18	2.16	0.60	0.14
120	2.50	0.21	2.59	0.74	2.16	0.38	0.14
150	2.00	0.17	2.07	0.60	2.16	0.30	0.14
180	1.67	0.14	1.73	0.50	2.16	0.24	0.14
200	1.50	0.12	1.55	0.44	2.16	0.22	0.14
210	1.43	0.12	1.48	0.42	2.16	0.22	0.14
240	1.25	0.10	1.29	0.38	2.16	0.18	0.14

Loss due to time- and frequency averaging as calculated using Eqs. (11.2) and (11.1) assuming a 1300 km baseline. Note that the expression given for frequency smearing is independent of observing frequency, only the channel bandwidth is important

central frequency ν_c since the synthesised beam also scales with ν_c, so that only the bandwidth is important.

For time smearing, we may use their formula 18–43, assuming a 12 h average over a circular UV-coverage with Gaussian tapering:

$$\frac{I}{I_0} = 1 - 1.22 \times 10^{-9} \left(\frac{r}{\theta}\right)^2 \Delta t^2 \qquad (11.2)$$

where Δt is the averaging time in seconds.

What loss to define as acceptable of course depends on your science, in particular the brightness of your target, but as a general guide one may tolerate 5% loss in amplitude due to averaging. Using the standard LOFAR raw data values, we have calculated the corresponding circle (diameter, to compare with station FWHM) for different observing frequencies, see Table 11.2. We have also included estimates for a typical long baseline observation averaging to 2s and 4ch/subband.

11.3 Calibration of International LOFAR Stations

In this section we describe, in general terms, how to calibrate phase and amplitudes of visibilites on baselines to international stations. Note that this strategy does also calibrate NL-stations, but using visibilites on baselines to international stations. More detailed information can be found in the LOFAR imaging cookbook.

11.3.1 Phase Calibration Using International Stations

Accurate phase calibration of the visibilities for a weak (or unknown) target source is usually done by finding residual phase errors using a calibrator (point-like or with a good model) close to the target source, and then transferring the derived corrections to the target. In principle, phase corrections can be determined separately for each channel if the data has high enough signal-to-noise. In practice, the requirement of a nearby calibrator often means that the calibrator is too weak, and therefore it is necessary to average in time and/or frequency to gain sufficient S/N to find the desired phase corrections. This can however only be done if there is no residual delay (causing a change of phase with respect to frequency) or rate (causing a change of phase with respect to time) present in the data. Unfortunately, residual delays and rates are common on international baselines, and we have to deal with them to be able to correct residual phase errors before imaging. Below we first describe the origin and timescales to be expected for delays and rates in LOFAR HBA data. Then we discuss how to use VLBI self-calibration techniques (i.e. fringe-fitting) to remove desidual rates and delays from the data.

11.3.1.1 Residual Delays and Rates

The phase of a single visibility depends on the time delay of the signal to reach two different stations. A given time delay will cause different phase-errors for different observing frequencies. We define the phase delay as $\tau_\phi = \phi/2\pi\nu$. We clearly see that a delay between two stations will produce a phase slope as function of frequency, which is also visible in the data, see e.g. Fig. 11.4a.

Residual delays can arise due to multiple effects which are not accurately modeled in the LOFAR correlator, for example:

- Errors in station positions or target source positions
- Instrumental effects (e.g. atomic clock drifts)
- Propagation through the ionosphere.

If the model applied during correlation was perfect, all stations would see a delay offset of zero for an isolated compact source, but deviations are produced by several factors.

Errors in station positions (and currently at a much lower level errors in the Earth orientation parameters) produce residual delays on the order of 75 ns, varying with a 24 h periodicity. These errors can be expected to be greatly reduced in the near future. Errors in the *a priori* centroid position (for example from low-frequency catalogues, with a typical error of a few arcseconds) and/or extended structure on subarcsecond scales, introduce an additional delay. The maximum baseline between an international station and the LOFAR core (the correlator reference point) is 1000 km (see Table 11.1); a positional error of $1.5''$ will lead to a residual delay of $\sim$20 ns on this baseline.

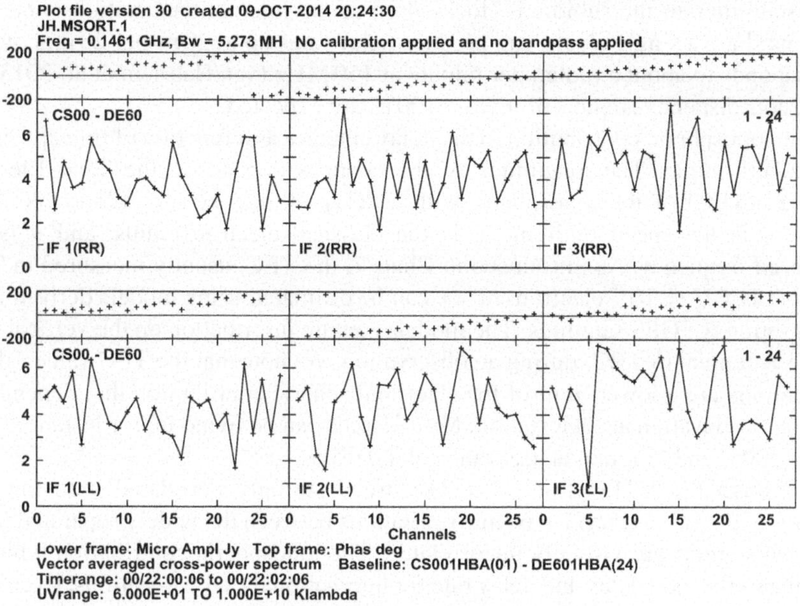

(a) Before fringe fitting

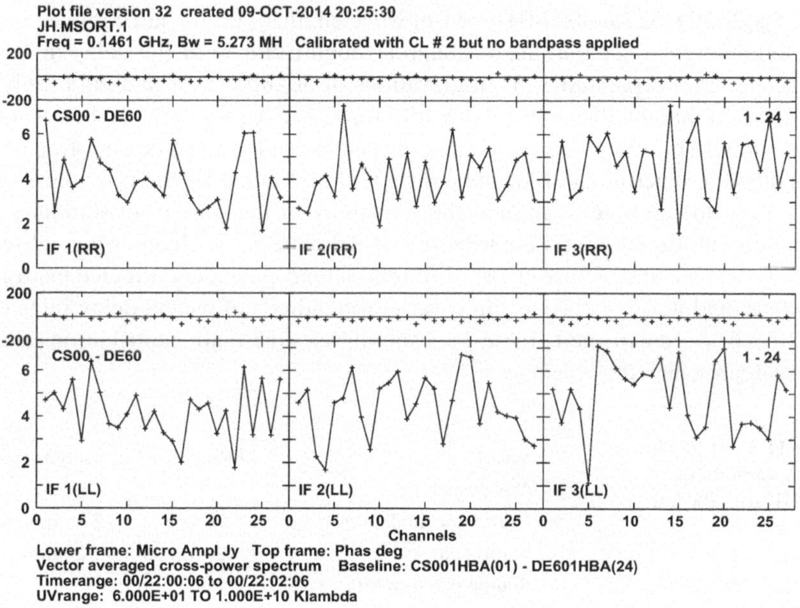

(b) After fringe fitting

Fig. 11.4 Two figures showing the effect of fringe fitting on 2 min of data on the baseline CS001HBA – DE601HBA for source J0958+6533 in project LC0_26. Both polarisations are shown, and the data are divided in three spectral windows (IFs in AIPS) of 5.3 MHz each. After applying the corrections from FRING, the phase is flat with respect to frequency, see (**b**), as it should be for a point source

Instabilities in the rubidium clocks used at the international stations (the core stations share a single clock) can produce delay rates up to 20 ns per 20 min, which corresponds to about a radian per minute at 150 MHz (van Haarlem et al. 2013). In total, non-dispersive delays of up to ∼100 ns are expected.

The ionospheric contribution to the delay changes as a function of time, position, and zenith angle. The magnitude of the changes depend on the Total Electron Content (TEC) of the ionosphere, with a delay of $\tau_{\rm ion} = c^2 r_{\rm e}/(2\pi\nu^2) \times$ TEC, where c is the speed of light, $r_{\rm e}$ is the classical electron radius, and ν is the observed frequency. As introduced in Chap. 7, the TEC, usually measured in TEC Units (1TECU $= 10^{16}$ electrons $\rm m^{-2}$), can be estimated using models derived from observations of GPS satellites. The models contain information on the vertical total electron content (VTEC) during an observation. We note that the TEC values above the stations are a lower limit of the slant ionospheric contribution that depends on the source elevation at each station. More details can be found in, for instance, Nigl et al. (2007) and Sotomayor-Beltran et al. (2013).

Although the VTEC follows a 24-h trend strongly correlated with the Sun elevation, the short-term (10–60 min) variations between the widely separated international stations are virtually uncorrelated. The ionospheric contribution typically dominates the total delay and delay rate for international LOFAR stations. Even after a complete phase calibration, the residual ionospheric delays between the calibrator and the target source can be important. We have used VLBI observations (VLBA project code BD152) at 300 MHz, or 1 m wavelength, of bright and compact pulsars at different angular separations to obtain a rough estimate of the delay difference between sources separated 1–5° at elevations of 50–80°. As a first approximation we estimated that the dispersive delay difference between sources at different lines of sight should be about 5 ns per degree of separation, for a source elevation of 60°.

Finally, the determination of the delays will be limited by the brightness of the source used to calibrate them, and the sensitivity of the station. In summary, in a normal observation the total contribution of delay to phase changes are caused by source position and structure errors, differential ionosphere, uncorrected instrumental delays, and noise. Delays of up to several hundreds of ns and delay rates of up to ∼20 ns $\rm h^{-1}$ are expected. Table 11.3 summarises the main contributions and the time scale in which they change.

Table 11.3 Approximate delay contributions at 140 MHz to a 700 km baseline

Effect	Delay	Time scale
Non-dispersive		
Correlator model error	∼75 ns	24 h (periodic)
Station clocks	∼20 ns	∼20 min
Source position offset (1.5″)	∼5 ns	–
Dispersive		
Slowly varying ionosphere	∼300 ns	∼hours
Rapidly varying ionosphere	$\gtrsim$10 ns	∼10 min
Differential ionosphere (source elevation 60 deg)	5 ns/deg sep.	–

11.3.1.2 Correcting Residual Delays and Rates

Due to the large and time-variable delay offsets at each station, solving for phase corrections directly (approximating the correction as constant over a given solution time and bandwidth) would require very narrow solution intervals for VLBI, and hence an extremely bright calibrator source ($\gtrsim$10 Jy). However, such a source would be unlikely to be close on the sky to the target. With a separation of perhaps tens of degrees the differential atmosphere/ionosphere between the calibrator and the target direction would render the derived calibration useless in the target direction. We can make use of fainter calibrators closer to the target with the VLBI phase calibrator known as *fringe-fitting*: simultaneously solving for three parameters (phase, non-dispersive delay, and phase rate) in each solution interval, allowing the solution duration and bandwidth to be greatly extended. This technique is very similar to ordinary phase calibration, but in addition to solving for phase we solve for a phase change with respect to frequency (delay) and time (rate) (Fig. 11.5).

At the time of writing, fringe-fitting (globally) solving for delays and rates is not available within common LOFAR software or in CASA, although in the latter case a task for this purpose is under development. The current best approach is to use the task *FRING* available within the Astronomical Image Processing System (AIPS[1]) (Greisen 2003). This task can however only solve for a *linear* change of phase with respect to frequency and time, i.e. a non-dispersive delay and constant rate within a specific solution interval.

Two options present themselves: to add additional parameters (covering dispersive delay and dispersive delay rate) to the global fit, or to reduce the solution bandwidth such that the constant dispersive delay approximation becomes valid again. The former option is obviously preferable from a sensitivity perspective, but greatly expands and complicates the solution search space. Efforts are underway to implement such an expanded fit, including in addition differential Faraday rotation, which becomes increasingly important at frequencies below 100 MHz. First tests on individual long baselines of LOFAR as well as baselines to other telescopes are promising, but the algorithms are not yet sufficiently mature for public use. Accordingly, we focus here on sources which can serve as primary calibrators under the latter set of conditions, where solution bandwidths are limited to no more than a few MHz and time intervals of a few minutes.

11.3.1.3 What is a Bright Enough Calibrator?

To derive delay and rate corrections we need to find a calibrator which is bright enough to find solutions in a small enough block in time and frequency so that the linear approximation is valid. The system equivalent flux density (SEFD) of a single

[1] http://www.aips.nrao.edu/index.shtml.

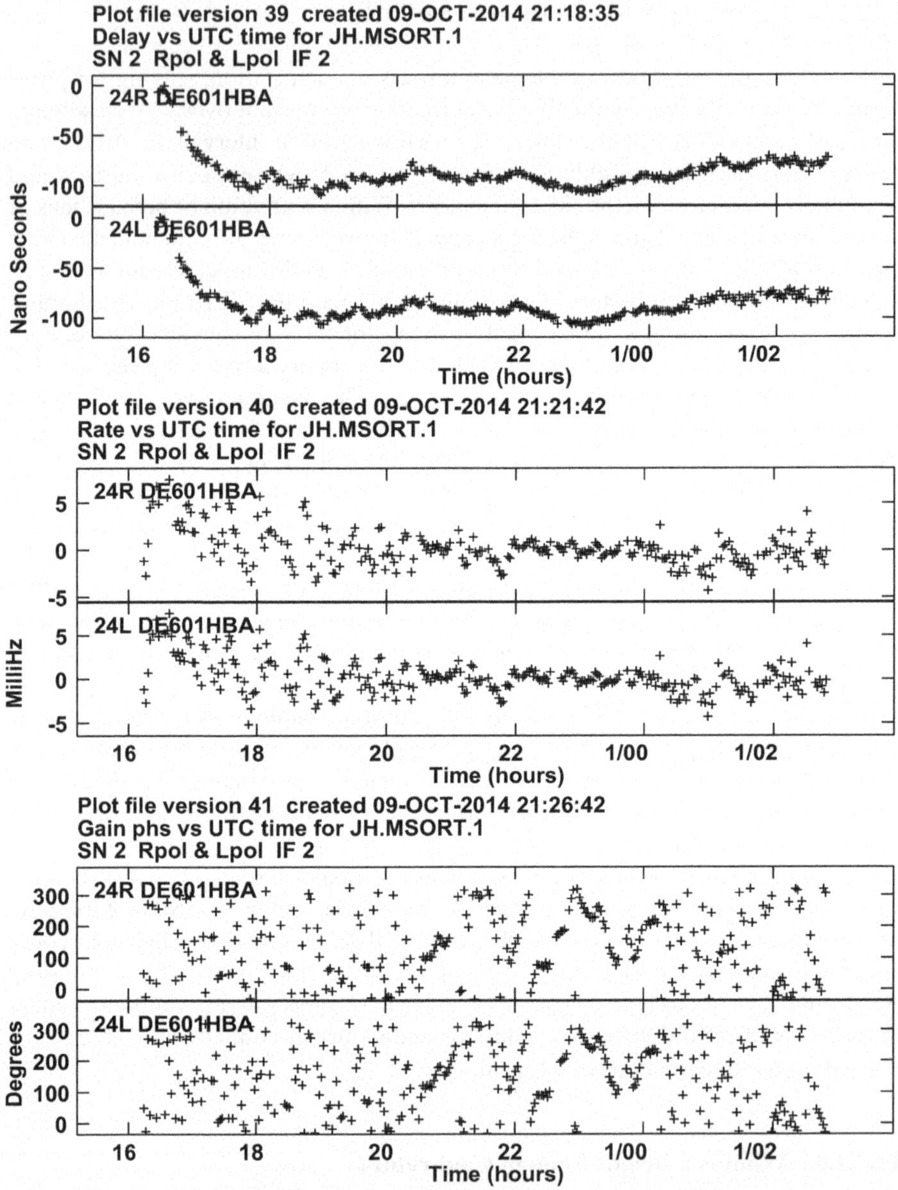

Fig. 11.5 Delay (top), rate (middle) and phase(bottom) corrections derived for the source J0958+6533 at 154 MHz by FRING for station DE601HBA for right and left circular polarisation. These plots show the corrections derived for a 10 h observation (the first segment of project LC0_026). It is clear from the rates and phases that phases changes rapidly during the first and last hours of the experiment. The delay solutions are more stable, although there is a large change during the first hour. In general, the ionosphere is more stable during midnight than at sunset or sunrise

LOFAR core station is approximately 1500 Jy[2] at a frequency of ~140 MHz (van Haarlem et al. 2013). An international station has twice the collecting area of a core station at ~140 MHz, so the expected SEFD is around 750 Jy. The theoretical 1σ baseline sensitivity of an international station to a (joined) HBA core station, given 3 MHz of bandwidth and 4 min of observing time, is 40 mJy in a single polarisation. If we require a minimum signal to noise ratio of 5 for fringe-fitting, this means we need a calibrator brighter than 200 mJy.

It is possible to use even weaker calibrators by forming a combined station of all the core stations, resulting in a very sensitive station with an SEFD of ~65 Jy (see Sect. 11.5.3). The theoretical 1σ baseline sensitivity of an international station to the phased-up core station, given 3 MHz of bandwidth and 4 min of observing time, is hence 8 mJy in a single polarisation. Given this increase in sensitivity, sources with flux densities as low as 50 mJy could in theory be used as delay/rate calibrators. However, in practice the sensitivity may be reduced by, for example, failing tiles, imperfect station calibration and correlated (astronomical) noise. Also, solution intervals shorter than 4 min could be required under suboptimal ionospheric conditions. Therefore, 50 mJy should be considered a lower limit, and in practice a flux density of more than 100 mJy at subarcsecond scales may be needed.

11.3.1.4 What is a Close Enough Calibrator?

In addition to being sufficiently bright, the delay/rate calibrator must be close enough to the target field that the differential delay between the two fields does not lead to decorrelation when phase-only secondary calibration is performed (as done for cm VLBI, either using the target itself or using another calibrator). The solution bandwidths are narrower by a factor of $\gtrsim$10 than for cm VLBI, which is helpful, but the ionospheric delay (inversely proportional to observing frequency squared) is much greater, meaning that on balance a calibrator closer than the $\lesssim$5° typical for cm VLBI will be needed. The maximum acceptable separation will be a strong function of ionospheric conditions and elevation, but at face value, given a bandwidth 20 times narrower (e.g., 3 vs 64 MHz) and frequency 10 times lower (140 vs 1400 MHz), one would expect that the calibrator would typically need to be separated by $\lesssim$1°. This is borne out by commissioning observations with LOFAR, which have shown acceptable results with separations up to several degrees in favourable ionospheric conditions, and unacceptable results with separations as small as ~0.8° in poor conditions. Ideally, then, a primary calibrator for International LOFAR observations would be located $\lesssim$1° from the secondary calibrator/target field to give acceptable calibration under most circumstances. Exploration of the distribution of compact sources at 140 MHz has shown that the density of calibrators on the sky could be enough to overcome this restriction (Moldón et al. 2015). This leads to

[2] A LOFAR core station consists of two sub-stations (2 × 24 tiles) in the HBA.

one calibration advantage of International LOFAR compared to cm VLBI; since the beam of an International LOFAR station is $\gtrsim 2°$ across, the calibrator may easily be observed simultaneously with the target source. Note however that care has to be taken when averaging the data to not reduce amplitudes severely by smearing, see Sect. 11.2.3.1. This can for example be avoided by using the *shift+averaging* procedure, see Sect. 11.5.1.2.

11.3.2 Amplitude Calibration of International Stations

In principle, instrumental gains within LOFAR could be tracked with time as done in cm-VLBI (this option is currently being commissioning with the COBALT correlator), but since this option is not yet available we rely on calibrators with known flux density.

For core and remote stations one can use, for example, standard flux density calibrators such as 3C196, or bright sources in low frequency catalogs such as MSSS (Heald et al. 2015). However, for international stations this is in general not possible because of the small spatial scales sampled by baselines to these stations. The bright standard flux calibrators, e.g. 3C196, have a very complex structure at subarcsecond scales. If we had a good model of this structure at our frequency of observation, we could in principle account for it in the calibration of the international stations. Work is being done to map well known flux density calibrators with high enough resolution, but until these models are available we have to rely on another bootstrapping technique.

The current best approach is to include two calibrators in the observations: a well monitored flux density calibrator (for instance 3C196 or 3C84) and another compact source with flux density of a few hundred mJy. The compact calibrator can often be the same as used to derive delay/rate solutions, see Sect. 11.3.1.3. While this compact source can be variable (on timescales longer than the observation), and hence not a suitable *absolute* flux density reference, it can be used to calibrate the *relative* amplitudes of all baselines, including international stations. The absolute flux density scale can then be set by fitting a common scaling factor for all visibilities (including the international baselines), by comparing the derived flux density for the standard flux density calibrator on short (NL) baselines, where subarcsecond scale structure is not important.

After a phase and amplitude calibration of an International LOFAR observation, the resulting visibilities should look similar to the ones shown in Fig. 11.6, where we show the calibrator J0958+6533 used in Varenius et al. (2015) to derive delay and rate corrections as well as the relative amplitude scale.

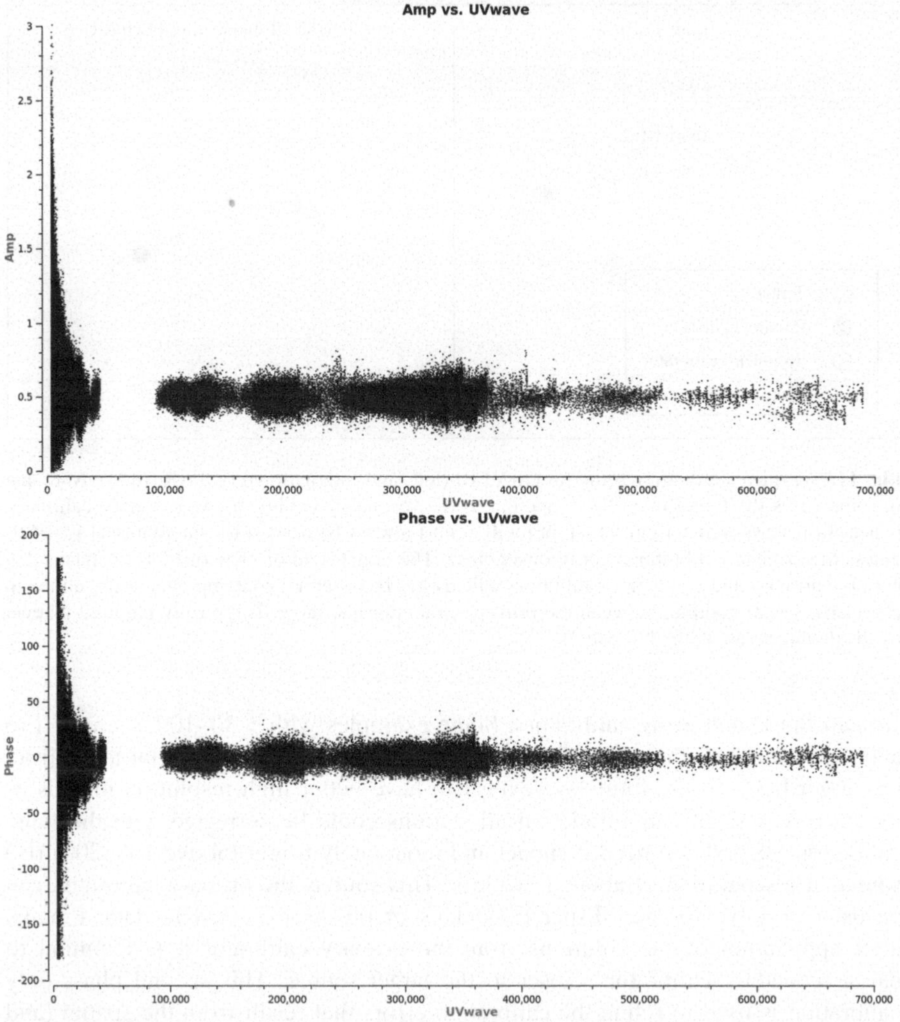

Fig. 11.6 Amplitude (top) and phase (bottom) vs *uv* distance for a calibrated dataset of J0958+6533. This calibrator has a very compact component of 0.5 Jy, mainly unresolved with the longest LOFAR baselines. We can see how the shorter baselines are sensitive to a much more complex structure and possibly other sources in the field

11.4 Observing Strategy

For a complete International LOFAR calibration we require different calibrators aimed to correct specific parts of the data. First, we need a very bright and well known source with a stable flux density that is used to set the flux scale. Although there are no good high resolution models for this kind of source, we can still find bright calibrators to set the flux scale of the short-baseline part of the array, which

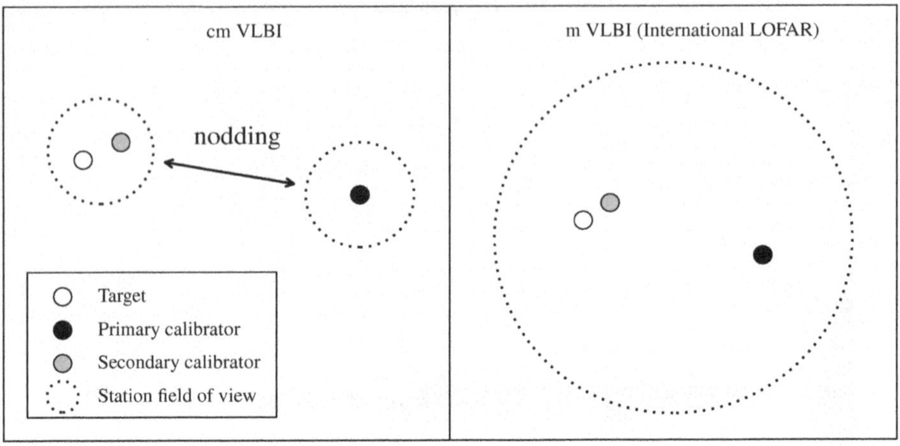

Fig. 11.7 Typical calibration setup for cm VLBI (left) and International LOFAR (right). Note that in some cases the target may itself function as the secondary calibrator. A secondary calibrator is not always required for cm VLBI, but will almost always be needed for International LOFAR, unless the primary calibrator is fortuitously close. The larger field of view of LOFAR means that both the primary and secondary calibrators will always be observed contemporaneously, unlike in cm VLBI, where nodding between the primary calibrator and target is typically required (shown by the double arrow in the left panel)

we call the Dutch array calibrators. Some examples with >30–100 Jy are 3C196 and 3C84, among others. The Dutch calibrator can be considerably far away from the target field. In the future we expect to have better high resolution models of these sources so the amplitudes of all stations could be corrected with the same calibrator. Second, we need compact and moderately bright (above 100–200 mJy) source at a separation of about 1° or less. This source, the primary calibrator, can be used to solve for non-dispersive delays of blocks of ∼3 MHz data. Finally, after application of the solutions from the primary calibrator it is common to use a secondary calibrator[3] closer to the target source. This second phase-only calibration is used to refine the calibration errors that result from the spatial (and temporal) interpolation of the primary solutions. Ideally, the secondary calibrator should be at ∼arcmin separations from the target source. Because solving for phases is a problem with fewer degrees of freedom than fringe fitting, lower S/N data can be used. Additionally, because the bulk delay has already been removed the residual delays should be small and even more bandwidth can be combined in a single solution for a further improvement in S/N. A secondary calibrator can therefore be considerably fainter (usually $\gtrsim$5 mJy). If bright enough, the target source can be used to derive this correction. This typical calibration strategy is illustrated in Fig. 11.7. In comparison with a typical cm-VLBI strategy a LOFAR

[3]A secondary calibrator is often referred to as an "in-beam" calibrator in VLBI if it is close enough to the target source to be observed contemporaneously.

observation can be more efficient because the size of the LOFAR station beam (see Table 11.2) is large enough that the primary and the secondary calibrators are observed simultaneously.

When preparing an International LOFAR observation one should take into account the different calibrators needed, considering that primary and secondary compact calibrators are currently unknown. Probably a small part of the observation should be aimed to identify calibrators. Moldón et al. (2015) proposed the following approach for an International LOFAR observation:

1. Identify candidate primary calibrators up to separations of a few degrees by using any of the criteria discussed in Sect. 5.3 in Moldón et al. (2015);
2. Conduct a short observation in snapshot mode as described in Sect. 11.5.2 before the science observation to identify the best primary calibrator (or calibrators).
3. If required and time permits, follow up with a "full bandwidth" snapshot observation to identify one or more secondary calibrators;
4. Set up the scientific observation to dwell on the field containing the primary calibrator and the target/secondary calibrator;
5. Include periodic scans (every $\sim$ hour) on a bright Dutch array calibrator to calibrate the core stations in order to form the tied station.
6. Shift phase centre to primary calibrator, preprocess and obtain delay solutions as described in this paper, apply them to the unshifted dataset;
7. If a secondary calibrator is to be used and is not yet identified, select 10 min of data and perform shift/averaging to candidate secondary calibrator sources;
8. If secondary calibrator is used: shift and average primary-calibrated dataset, image and selfcalibrate, apply solutions to the unshifted dataset;
9. Shift and average calibrated dataset, image and (if needed) selfcalibrate target.

In the near future, the pipeline used for this project will be developed, in collaboration with the LOFAR operations team, into an expanded form capable of carrying out the approach described above. This pipeline will be made available to all International LOFAR observers, delivering a reduced data volume for long-baseline observations and enabling calibrated data to be more quickly produced.

11.5 Practical Considerations

The calibration of International LOFAR data is not fundamentally different from any other radio interferometric data, although special care should be taken regarding, for example, the phase calibration using VLBI techniques. However, there are some particular practical considerations that could help the observer to reduce this type of data. A detailed description of the steps and software that can be used to reduce these data can be found in The LOFAR Imaging Cookbook.[4] In this section we

[4]http://www.astron.nl/radio-observatory/lofar/lofar-imaging-cookbook.

list some hints and useful procedures that are needed when planning and reducing International LOFAR data.

11.5.1 Optimising Use of Available Bandwidth

It is possible to distribute LOFAR bandwidth over a number of beams to simultaneously observe different regions of the sky. In particular, it is possible to divide the bandwidth on target(s) and calibrator, which provides a continuous source calibration without the need of regularly nodding from target to calibrator. Another possibility is to distribute the bandwidth among a large number of sources to search for suitable calibrators. For example one can generate 30 beams to observe simultaneously 30 sources with 3 MHz bandwith as a fast way to search for suitable compact calibrators for an International LOFAR observation (see i.e. Moldón et al. 2015).

11.5.1.1 Unequal Distribution of Subbands on Target and Calibrators

If your calibrator is bright, you can use fewer subbands on the calibrator, and thereby get better sensitivity on the target. To use fringe finding, we need to sample accurately the residual delay/rate slope (and possibly curvature at low frequencies) present in the data. This can be done with sparse sampling in frequency, where the optimal coverage is achieved by spreading the subbands as a powerlaw density with denser placement of subbands at lower frequencies (Martí-Vidal 2010). The advantage of this approach is that more bandwidth can be placed on the target. The disadvantage is that the calibration becomes a bit more demanding. One reason for this is that the UVFITS format used by AIPS (for running fringe fitting) requires data in all channels. If we do not have contiguous subband coverage in frequency, we need to insert fake data and flag that (e.g. using NDPPP) before reading the data into AIPS. This will cause an increase in data volume which will slow down processing. Also, spreading the subbands sparsely is always a risk in case your calibrator is weaker than you think. A detailed discussion can be found in Martí-Vidal (2010), where the authors analyse how to distribute subbands specifically for LOFAR observations for optimal fringe detection.

11.5.1.2 The Shift + Averaging Procedure

Given the high resolution obtained with the International LOFAR observations, imaging of the region restricted by the time and frequency average of the data (see Table 11.2) would require a very high computational cost. If one is interested in multiple objects within the station beam, one can phase-shift (and re-project) the *uv*-data to each object before averaging. After correlation, the full-resolution visibility dataset can be shifted and averaged multiple times, to the positions of all the target sources and possibly to one or more nearby calibrators. In the future, it will be possible to request shifting and averaging of data to multiple phase centres within a beam as a part of a normal observation.

11.5.2 Finding Calibrators

Until a good catalogue of compact sources at MHz frequencies is available, it is important to take into account that a science observation might require a preparatory search of calibrators. A fast method using the distribution of bandwidth between many sources (see Sect. 11.5.1.1) is described in Moldón et al. (2015). A pre-selection based on a number of parameters from existing catalogues, such as the low-frequency spectral index, and the flux, can be performed to optimize the search. In particular the most useful catalogues are the VLSSr, at 74 MHz, 4 m wavelength (Lane et al. 2012), the WENSS catalogue at 325 MHz, 92 cm wavelength (Rengelink et al. 1997), the TGSS-ADR1, at 151 MHz, 2 m wavelength (Intema et al. 2017), and especially the Multifrequency Snapshot Sky Survey (MSSS), which comes from LOFAR observations (Heald et al. 2015). Also, compact calibrators at mas scales detected with cm-VLBI are probably also compact at LOFAR frequencies. Although there is a strong correlation, a cm-VLBI calibrator may not be suitable for LOFAR in case it is variable or if it has inverted or gigahertz peaked spectra. An updated list of cm-VLBI calibrators covering the whole sky can be found in http://astrogeo.org/rfc/.

Moldón et al. (2015) showed that there is a density of ~1 good calibrator per square degree based on two fields with Galactic latitudes of +26.6° and +43.4°. However, we expect less compact sources at lower Galactic latitudes due to interstellar scattering. The Galactic electron density model NE2001 (Cordes and Lazio 2002) predicts an scattering at a galactic latitude of 50° of almost 100 mas at 150 MHz, which is five times smaller than our resolution. However, the scattering is about 300 mas, similar to our beamsize, at latitudes of 5–10°, depending on the longitude. Therefore, observations below a Galactic latitude of 10° are likely to be affected by scattering on the longest baselines, and the effect should be severe below about 2°, especially towards the Galactic Center.

The ongoing LOFAR Long-Baseline Calibrator Survey (LBCS; Jackson et al. 2016) has the aim of identifying a network of sources over the entire northern sky that are suitable as calibrators for the LOFAR international baselines. The LBCS database is publicly available online[5] and provides the growing list of calibrators.

[5] http://vo.astron.nl/lbcs/lobos/cone/form.

11.5.3 Form a Combined Station

When studying very compact structures, the shortest baselines do not add much interesting information while they slow down the calibration process. The core stations can be added to form a coherent "tied station" (TS001) that keeps the total core sensitivity on the long baselines to the international stations. Since the core stations are under similar atmospheric conditions and they share the same clock only slow changes in their amplitudes and phases are expected, and thus they can be calibrated by observing a bright primary calibrator once every ~1 hr. TS001 can be formed by summing baseline visibilities with the NDPPP task "StationAdder". After this step, all original visibilities with core-core baselines can be discarded using the NDPPP task "Filter" to significantly reduce the data volume.

One important benefit of having a tied station is that it works as a very sensitive station. This tied-array station aids in the derivation of calibration solutions to the international stations with FRING, and can be used as a reference station. We note that a tied-station formed by adding the whole core has a very small (5% amplitude loss at $30''$ distance from phase centre) field of view. Although this is rarely a problem for deriving FRING solutions, care is needed if using such combined data to image extended objects.

11.5.4 Getting LOFAR Data into AIPS

The task FRING in AIPS can be used to remove residual delays and rates in the data. However, AIPS requires that the data are in circular polarisation (LOFAR usually stores data in linear polarisation). AIPS also requires the data to be converted from measurement set (MS) to the UVFITS file format. In this section we describe how this can be achieved.

11.5.4.1 Converting Linear to Circular Polarisation

Differential Faraday rotation (see Chaps. 7 and 10) introduces rapid phase changes with frequency into linear polarisation data on long baselines. For long-baseline observations is preferable to work in a circular (R,L) polarisation basis. In this basis, the ionospheric disturbances are transformed from coupled amplitude/phase effects (as in the linear X,Y basis) to phase only effects. Since differential Faraday rotation does not mix R and L polarisations we may calibrate RR and LL independently. Furthermore, standard VLBI techniques like fringe fitting work in a circular (R,L) polarisation basis. To run FRING in AIPS, the data have to be converted to circular polarisation.

There are two common ways to convert LOFAR measurement sets to circular polarisation: either using BBS and Table Query Language (TAQL), which uses the BBS beam model, or using the custom software *mscorpol* developed by T. D. Carozzi, which uses its own beam model. Detailed instructions on how to use these tools are provided in the LOFAR Imaging cookbook.

11.5.4.2 Converting Measurement Set to UVFITS

Since AIPS understands the UVFITS-format, but not Measurement Sets (MS) we need to convert the data from MS to UVFITS. This can be done using e.g. the tool *ms2uvfits* available at the LOFAR cluster or the task *exportuvfits* in CASA. Note that it is important to have contiguous data in frequency (e.g. by filling missing subbands with fake data) and to have data present for all baselines (i.e. by flagging instead of partially filtering antennas).

Acknowledgements This chapter and all the techniques and procedures described in it are based on the work and many useful contributions of the LOFAR long baselines working group. AIPS is produced and maintained by the National Radio Astronomy Observatory, a facility of the National Science Foundation operated under cooperative agreement by Associated Universities, Inc.

References

Cordes, J.M., Lazio, T.J.W.: ArXiv Astrophysics e-prints (2002)
Greisen, E.W.: Information Handling in Astronomy - Historical Vistas, vol. 285, p. 109. Kluwer, Dordrecht (2003)
Heald, G.H., Pizzo, R.F., Orrú, E., et al.: Astron. Astrophys. **582**, A123 (2015)
Intema, H.T., Jagannathan, P., Mooley, K.P., Frail, D.A.: Astron. Astrophys. **598**, A78 (2017)
Jackson, N., Tagore, A., Deller, A., et al.: Astron. Astrophys. **595**, A86 (2016)
Lane, W.M., Cotton, W.D., Helmboldt, J.F., Kassim, N.E.: Radio Sci. **47**, 0 (2012)
Martí-Vidal, I.: Astron. Astrophys. **517**, A83 (2010)
Moldón, J., Deller, A.T., Wucknitz, O., et al.: Astron. Astrophys. **574**, A73 (2015)
Nigl, A., Zarka, P., Kuijpers, J., et al.: Astron. Astrophys. **471**, 1099 (2007)
Rengelink, R.B., Tang, Y., de Bruyn, A.G., et al.: Astron. Astrophys. Suppl. Ser. **124**, 259 (1997)
Sotomayor-Beltran, C., Sobey, C., Hessels, J.W.T., et al.: Astron. Astrophys. **552**, A58 (2013)
Taylor, G.B., Carilli, C.L., Perley, R.A. (eds.): Synthesis Imaging in Radio Astronomy II. Astronomical Society of the Pacific Conference Series, vol. 180 (1999)
van Haarlem, M.P., Wise, M.W., Gunst, A.W., et al.: Astron. Astrophys. **556**, A2 (2013)
Varenius, E., Conway, J.E., Martí-Vidal, I., et al.: Astron. Astrophys. **574**, A114 (2015)

Chapter 12
Spectral Line Data Analysis with LOFAR

J. B. Raymond Oonk

Abstract The Low Frequency Array (LOFAR) represents a major step forward in exploring the low-frequency radio regime. With its unprecedented sensitivity, spatial and spectral resolution, frequency coverage and multi-beaming capability, LOFAR enables efficient surveys of the sky that will revolutionize the field of low-frequency spectral line studies. In this chapter we present a very brief overview of line spectroscopy at radio frequencies below 250 MHz. We focus our discussion on (Galactic) radio recombination lines from carbon that are prominently visible at these low frequencies and that provide an important tracer of the physical properties of the cold neutral medium. We briefly discuss the technical capabilities and calibration procedures of spectral line data as observed with LOFAR.

12.1 Introduction

The low-frequency (<250 MHz) radio regime is a relatively unexplored region for spectral lines. The lack of exploration of this frequency range has been largely due to the absence of sensitive low-frequency telescopes in combination with high spatial and spectral resolution as well as wide frequency coverage. This situation is now changing with the advent of new, powerful, low-frequency radio telescopes such as the Low Frequency Array (LOFAR; van Haarlem et al. 2013), the Murchison Widefield Array (MWA; Tingay et al. 2013) and the Long Wavelength Array (LWA; Ellingson et al. 2013). Here we will focus on the spectral line capabilities of LOFAR that is able to observe in the 10–250 MHz window.

The low-frequency domain contains a wealth of astronomically important spectral line transitions. An overview of which lines are present at frequencies below

J. B. R. Oonk (✉)
Netherlands Institute for Radio Astronomy (ASTRON), Dwingeloo, The Netherlands

Leiden Observatory, Leiden University, Leiden, The Netherlands
e-mail: oonk@astron.nl

G. Heald et al. (eds.), *Low Frequency Radio Astronomy and the LOFAR Observatory*, Astrophysics and Space Science Library 426,
https://doi.org/10.1007/978-3-319-23434-2_12

250 MHz can be obtained from, for example, the splatalogue catalog.[1] In the following we will focus on radio recombination lines (RRL), as currently these are the most prominent spectral lines at these frequencies.

However, other spectral lines of importance, that are yet to be detected, are for example the nitrogen (N I) hyperfine splitting transition and molecular lines (e.g. from NO, H_2CO and OH). Finally there is also the possibility of observing higher frequency lines that for objects at large redshifts (z) are shifted into the observable range for LOFAR. Examples of such lines are the hydrogen (H I) 21 cm hyperfine splitting line at 1420 MHz ($z \geq 4.7$), the deuterium (D I) line at 327 MHz ($z \geq 0.3$) and the OH 1612, 1665, 1667, 1721 MHz ($z \gtrsim 5.9$) lines.

12.1.1 Radio Recombination Lines

Spectral lines resulting from atoms recombining with electrons in a diffuse, ionized plasma are important diagnostics probing the conditions of the ISM. At radio frequencies we refer to these lines as radio recombination lines (RRL); see e.g. Gordon and Sorochenko (2009); Peters et al. (2011) and Oonk et al. (2015) for recent reviews. RRLs can broadly be separated into two classes: discrete and diffuse. Discrete RRLs from hydrogen (HRRL), helium (HeRRL) and carbon (CRRL) are typically observed above about 1 GHz. Here the RRL spectrum is dominated by HRRL and HeRRL lines that are found to be excellent tracers of dense, warm, ionized gas.

At lower frequencies, i.e. below about 1 GHz, we predominantly observe diffuse RRLs from carbon and hydrogen. Diffuse RRLs are thought to arise in cold, atomic clouds. The observed characteristics of these RRLs (i.e. the integrated optical depth and line width) depend sensitively on the conditions of the emitting gas. By combining RRL measurements with appropriate non-LTE[2] models it is therefore possible to obtain the physical properties (i.e. temperature and density) of the emitting gas (e.g. Salgado et al. 2017; Oonk et al. 2015, and references therein). Low-frequency RRLs therefore provide a unique view of the cold neutral medium (CNM) and LOFAR is ideally suited to observe these lines.

12.2 Spectral Capabilities of LOFAR

LOFAR has three combined antenna and filter setups that allow for observing in the ranges 10–90 MHz with the Low Band Antenna (LBA), 100–200 MHz with the High Band Antenna and the LOW filter (HBA), and 200–250 MHz with High Band Antenna and the HIGH filter (HBA-HIGH). LOFAR furthermore has a very

[1] http://www.cv.nrao.edu/php/splat/.

[2] Local Thermodynamic Equilibrium.

powerful and flexible software correlator as its backend that allows the user to observe up to 96 MHz of instantaneous bandwidth that is subsampled into 488 subbands. A subband is the most elementary block of LOFAR data and consist of chunk of frequency with a fixed width of 0.1953125 MHz. This subband width corresponds to about 1000 km/s at 60 MHz and 250 km/s at 240 MHz.

The central frequency of each subband is fixed by the clock. It is therefore not possible for the observer to alter the central frequency of a subband and tailor the frequency setup of an observation to a specific spectral line. Currently only the 200 MHz clock is available and thus the frequencies and frequency ranges available for all LOFAR observations are the same and fixed. The frequency ranges of the subbands do not overlap and therefore, as we will discuss below, there are small gaps between the subbands due to the bandpass roll-off at the outer edge channels of each subband. About 10 percent of the total bandwidth is lost to these bandpass effects and if a line of interest happens to falls in one the gaps between the subbands then it will not be possible to change the spectral setup in such a way that this line can be recovered. In the future the 160 MHz clock may become more widely used, and that to a large extent would allow the observer to fill in the gaps left by the 200 MHz clock.

The frequency resolution of LOFAR depends on the number of channels per subband. So far 64, 128, 256, 512 and 2048 channels per subband have been successfully used. This corresponds to resolutions varying from 0.5 to 15 km/s at 60 MHz and 0.1 to 4 km/s at 240 MHz. The channel frequencies reported in the measurement set for a subband correspond to the center of each channel. Observations with the full 96 MHz bandwidth and sampled with up to 256 channels per subband have been successfully carried out with LOFAR. However, for even higher spectral resolution the total data rate entering the correlator determines the combination of bandwidth and spectral resolution that can be observed.

12.3 Spectral Line Calibration

As an example of how to reduce and calibrate LOFAR spectral line data we have chosen to focus on a interferometric LBA observation of narrow, Galactic RRL lines in absorption towards the Galactic supernova remnant Cassiopeia A (Cas A). This 10 h observation was carried out with the full Dutch LOFAR array using 512 channels per subband and we observed simultaneously with two beams having the same frequency setup. The first beam was placed on Cas A, which is our target, and the second beam was placed on Cygnus A (Cyg A) which serves as our flux calibrator.

To illustrate some of the details involved with spectral line calibration we will focus our discussion on the reduction of a single subband containing a CRRL α line transition, in absorption, at 33.47 MHz (corresponding to a quantum level $n = 581$) and also in absorption a CRRL β line transition at 33.42 MHz (corresponding to a quantum level $n = 732$), see Fig. 12.1. This type of observation

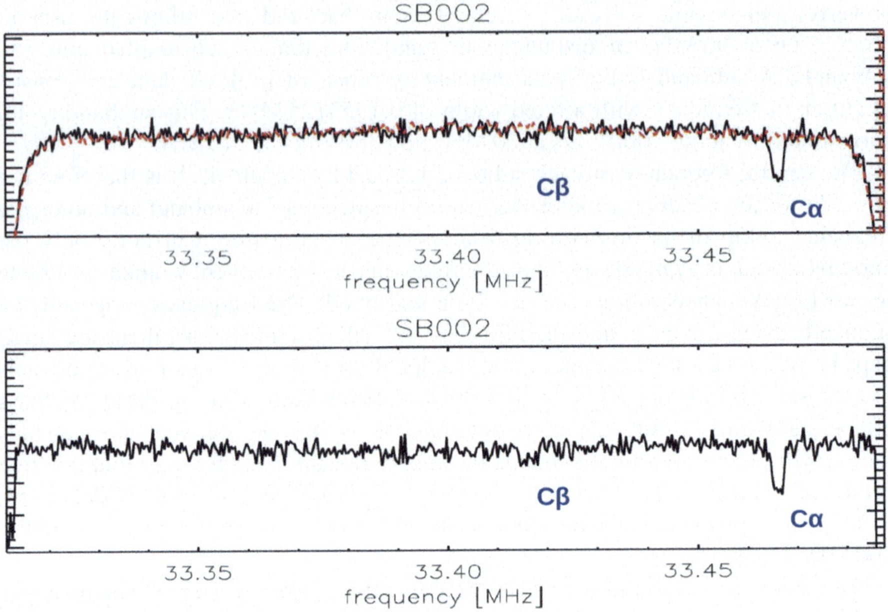

Fig. 12.1 LOFAR Low Band Antenna 10 h single subband spectra of Cas A and Cyg A at 33 MHz. The subband spectra are sampled with 512 channels and are part of the LOFAR Cassiopeia A Spectral Survey (LCASS; Oonk et al. 2017). (*Top*) In black (solid line) we show the Cas A spectrum and in red (dashed line) we show the corresponding Cyg A spectrum. Here we have already removed the spectral slopes across the subbands that correspond to the instrumental response of LOFAR in combination with spectral energy distribution of the two sources. The removal of these slopes was done by first blanking the CRRL absorption at 33.42 and 33.47 MHz, and then fitting a linear function to the central 80 percent portion of the subband. Note the very good agreement between the two spectra, except there where, as expected, the CRRL absorption lines are found. (*Bottom*) This shows the Cas A bandpass corrected spectrum upon dividing the Cas A spectrum by the Cyg A spectrum. Note that part of the bandpass drop-off in the outer edge channels can be recovered in this manner

represents a relatively simple and straightforward case where we have apriori a good understanding of the line frequency and its expected amplitude and width.

In general the reduction and calibration of spectral line data closely follows that of continuum data (see Chaps. 5 and 9). However, one of the major issues of dealing with high resolution spectral line data from LOFAR is the very large data sizes involved. The first step in the calibration process is to flag and, in the case of the LBA, demix the subband data for both our target and our calibrator. For flagging the AOflagger (Offringa 2010) is used. Both flagging and demixing is done within the NDPPP framework. In the case of demixing spectral line data we compute only one frequency averaged solution per time sample and then subtract these solutions from the frequency unaveraged data sets.

Next, using NDPPP, we create a frequency averaged (to one or a few channels per subband) copy of the flagged and demixed data set for Cyg A that will be used for continuum calibration. Here we have assumed that Cyg A does not contain strong line emission within the subband. If strong line emission is present then it is necessary to consider only those channels that are free of line emission when creating a frequency averaged data set for continuum calibration. We then calibrate this continuum data set of Cyg A using BBS (Pandey et al. 2009). The solutions, after careful inspection, are then applied to the frequency unaveraged Cas A data set. It is useful to not only transfer the amplitudes, but also the phases from the calibrator to the target, even if they are separated over distances larger than the isoplanatic patch, because to first order it is found that this removes uncertainties in the clock delays (see also related discussion in Chap. 7).

We now have a frequency unaveraged data set of Cas A that has been flux calibrated corresponding to the average over the subband bandwidth. However, the flux calibration as function of frequency, i.e. the bandpass, has not yet been obtained. This calibration is sufficient for our purposes because, (i) we are interested in narrow lines, (ii) the instrumental bandpass over the central 80 percent of the subband is to a good approximation linear, and (iii) radio continuum spectra are to a good approximation powerlaws and thus featureless across a subband.

In order to obtain the full bandpass solution one would need to perform a full calibration on the frequency unaveraged data set of the calibrator. For the LBA the standard strategy is to observe the calibrator and the target simultaneously. Therefore the data sets can become very large, i.e. a single subband for a 10 h observation with 512 channels could be as large as ~0.5 Tb. Calibrating the unaveraged subband data, although possible, would require a long time and significant computing resources thus making this procedure inefficient for large data sets containing many subbands with spectral lines.

The flux calibrated data set of Cas A still contains residual phase errors. It has been found that these residual phase errors can significantly affect the spectra, even when only considering single subbands. To correct these errors we perform phase selfcalibration (see Chap. 5). To make this procedure more efficient we create a frequency averaged copy of the flux calibrated Cas A data set for this purpose. In the case of strong line emission within the subband it is again necessary to consider only those channels that are free of line emission when creating this frequency averaged continuum data set for the target.

The final selfcalibrated phase solutions are then applied to the flux calibrated, frequency unaveraged Cas A data set and we are now ready to create a spectral line cube by imaging the data with AWimager (Tasse et al. 2013, and see Chapter 8). The size of this data set again means that imaging the full data set at once to directly create a 3-dimensional spectral line cube requires very large computing resources. To make this process more efficient we first split the data set into single channel data sets, i.e. in this 512 channel data set, we first split off each channel and image each of these individually. After convolving all channel images to the same spatial resolution we create the final spectral line cube by placing the convolved 2-dimension channel images into a 3-dimensional cube and attach to this cube the appropriate meta data for a measurement set and/or fits header in the case of a fits cube.

There is no dedicated cube viewer and/or analysis software for LOFAR. However, spectral line cubes can be analyzed by a variety of freely available data analysis and visualization packages such as for example *CASA* (McMullin et al. 2007) and *AIPS* (Greisen 2003). Spectral products, e.g. 2-dimensional line intensity, velocity and velocity dispersion maps, can be derived from the cube by these software packages and we leave it to the user to chose his or her favorite. Below we discuss in bit more detail a few additional aspects of the analysis that are important in order to derive science quality spectral products from calibrated LOFAR cubes.

12.3.1 Correcting the Bandpass

The intrinsic bandpass of LOFAR is given by the instrument response function. There are several contributions to this bandpass as discussed in van Haarlem et al. (2013) and Romein (2008). Here we will not discuss the broad frequency response, i.e. on scales larger than the width of a subband, of LOFAR and which is primarily determined by the antenna design and the front-end electronics. Instead we will briefly describe the effects of the polyphase filters (PPFs) that affect the bandpass across a single subband. The LOFAR data flow contains two stages of PPFs. The first stage PPF is located at the station level and subdivides the observed bandwidth in to subbands with widths of 0.1953125 MHz. The second stage PPF, located at the correlator level, splits each subband into channels. These PPFs affect the bandpass by introducing a wave pattern across a subband as a function of frequency. This wave pattern results from a non-uniform response of the PPF. However, this response has been computed analytically and is removed at the correlator level (see Figs. 5 & 7 in Romein 2008, and Chapter 2 in this volume).

For any observation the actual observed bandpass is a combination of the instrument response function and the observed spectral flux distribution on the sky. If one observes an astronomical source on the sky with a well known spectral energy distribution, i.e. a flux calibrator, then it is possible to determine this intrinsic response function by comparing the observed raw data with the flux calibrated data. Normally, instead of determining the instrument response function, the flux calibrator is used to remove the imprint of the instrumental response function, as a function of frequency, from the raw data and thus obtain flux calibrated data.

As mentioned in Sect. 12.3 the LOFAR software stack does currently not contain a specific task to perform a bandpass correction on the data. In the example of Sect. 12.3 we carried out a frequency averaged flux calibration for a subband. This frequency averaged approach was taken in order to lower the computational cost and provides an amplitude correction corresponding to the weighted mean of the observed flux in a subband. For narrow lines, i.e. much narrower than the width of subband, that are situated within the central 80 percent of a subband this type of frequency independent flux calibration is sufficient.

For the calibrated and imaged spectral line cube of Cas A, in Sect. 12.3, containing narrow lines in the central portion of the subband we now have two options for correcting the bandpass. If the line frequencies are known we can determine the spectral bandpass over the subband by first blanking the line channels and then fitting a linear function (or low-order polynomial) to the remaining continuum channels. Dividing the subband by the fit then removes the bandpass. In this procedure it is useful to discard the outermost ~10 percent of the edge channels in a subband as these have a significantly reduced sensitivity.

The second option for narrow lines is to also create a spectral line cube for the calibrator in the same way and at the same frequency as was done for the target. From this cube it possible to derive a calibrator subband spectrum to perform a bandpass correction. Since flux calibration was done on the frequency averaged data the slope of the target and the calibrator subband spectra will differ due to the spectral index of sources and the instrument response at different elevations. To a good approximation, upon discarding the edge channels, this difference in slope can be removed by a linear function or a low order polynomial. As an example we show in Fig. 12.1 the subband spectra, after removing the slopes, for Cyg A (scaled to the flux level of Cas A) and Cas A for the example discussed in Sect. 12.3. It is evident that there is a good match between the calibrator spectrum (Cyg A) and the target spectrum (Cas A) and the bottom part of Fig. 12.1 shows that the bandpass correction determined in this manner is excellent.

LOFAR has digital filters with a very flat response in the central 80 percent of subband and therefore both of the options discussed above provide a valid approach for the bandpass correction of narrow lines. However, best results are often obtained by using an astronomical calibrator, i.e. the second approach, for computing and correcting bandpass. In the case of broad lines, i.e. line widths corresponding to a significant fraction of the subband width or more, the bandpass needs to be obtained by first combining several subbands in frequency for both the target and the calibrator. This is to ensure sufficient line and continuum coverage. Prior to combining the subbands in frequency the outer edge channels should be blanked.

Above we have outlined two approaches that provide a good bandpass correction in a computational efficient manner for large spectral line data sets containing many lines. However, in order to obtain the most accurate bandpass results, whether it is for narrow or broad lines, one needs to consider performing a full, frequency unaveraged *BBS* calibration of the calibrator.

12.3.2 Continuum Subtraction

To determine the line intensity or optical depth we need to estimate the continuum emission. For spectral line cubes this can be done either in the UV-plane or in the image plane. There is currently no dedicated LOFAR software available for doing this. Software packages such as *CASA* and *AIPS* contain tasks that are able to remove the continuum emission from a cube. In the case of *AIPS* the tasks *IMLIN* and *UVLIN* provide the necessary functionality. In the case of *CASA* the corresponding tasks are *imcontsub* and *uvcontsub*. In the case described in Sect. 12.3 we have used dedicated *IDL* procedures, similar to *IMLIN* and *imcontsub*, to remove the continuum in the image plane by fitting the continuum with a linear function. Upon fitting the continuum we discarded the outer edge channels that are affected by the bandpass roll-off.

In the case of RRLs the natural quantity for comparing the measurements to the models is the line optical depth (e.g. Salgado et al. 2017; Shaver 1975; Walmsley and Watson 1982). Once the continuum and the line intensity each have been estimated the line intensity can be converted to an optical depth by dividing it by the continuum.

12.3.3 Doppler Correction

Currently LOFAR does not support Doppler tracking during observations. This means that spectral lines are shifted due to the motion of the earth relative to the observed position on the sky. Doppler corrections for the motion of the Sun through the Galaxy and of the Earth around the Sun are both approximately constant over the duration of a typical 8 h LOFAR observation and thus can be corrected by applying a single shift in velocity from the observed frame to the local standard of rest. However, the Earths rotation around its own axis is not constant during the observation and this motion leads to velocity shifts less than 0.5 km/s.

There are several software packages that contain tasks that can correct line shifts in the recorded offline data when Doppler tracking was not performed by the telescope. Examples are the tasks *CVEL* in *AIPS* and *CLEAN* in *CASA*. LOFAR data is compatible with *CASA* and can be converted to become compatible with *AIPS*. The absence of Doppler tracking therefore does not affect our ability to detect and analyze spectral line data.

12.3.4 Flagging Statistics

Radio observations are affected by radio frequency interference (RFI). A quantitative estimate of the RFI situation for LOFAR has been presented in Offringa et al. (2013). These authors show that for LOFAR the RFI typically affects less than 5 percent of the data in either the 30–78 MHz range or the 115–163 MHz range, when observed with 64 channels per subband. Outside the frequency ranges studied by Offringa et al. (2013) the RFI percentage increases.

The RFI in the observable range for LOFAR is typically found to be very narrow and increasing the frequency resolution helps to decreases the RFI percentage. Nonetheless at frequencies below 20 MHz the RFI percentage can be very high. RFI fractions between 45 and 65 percent have been reported at 18 MHz (Salas et al. 2017). For sufficiently strong sources it remains possible to perform and detect spectral lines. However, for these high RFI fractions the associated flagging percentages can affect the quality of the spectral bandpass. Therefore it is important to keep track of the flagging statistics for each channel within a subband and carefully cross-check these against possible line features. This is especially important when the flagging statistics vary significantly on frequency scales similar to expected line widths (Salas et al. 2017).

12.3.5 Processing and Storage Requirements

A typical 8 h LOFAR observation with 38 stations (e.g. the Dutch array), 64 channels per subband and 1 s time sampling generates about 43 Gb (or 114 Gb in the case of a HBA dual observation) per subband. For the procedure outlined above in Sect. 12.3 a small compute cluster with 64 Gb memory and a single node with six CPUs can fully reduce, calibrate and image a single subband in about 6 h. In this time estimate we have not included the self-calibration process which can significantly increase the required computing time by factors of a few or more. For higher spectral resolution observations the required computing time scales in a roughly linear fashion with the number channels, if the available memory on the compute cluster is scaled up in a similar fashion.

12.4 Concluding Remarks

In this chapter we have provided a brief discussion of spectral line data analysis for LOFAR from its observing capabilities, through the reduction, calibration and imaging process. We have highlighted some of the LOFAR specific aspects in this process that otherwise is very similar to the standard procedures followed for many other radio telescopes. Finally we briefly discussed how to derive spectral line products from the calibrated LOFAR cubes. We conclude that LOFAR through its flexible and innovative design is a great instrument for spectral line observations that in the coming years, through e.g. low-frequency RRL observations, will greatly add to our understanding of the interstellar medium.

Acknowledgements A lot of work by many people has gone into characterizing LOFAR for spectral line observations. JBRO would like to thank the Radio Observatory, the Software Group and the Astronomy Group at ASTRON for enabling and supporting these observations. In particular JBRO has had many fruitful discussions with G. Heald, J. Hessels and the Pulsar group, J. McKean, J. D. Mol and J. Romein on the topics addressed in this chapter. A special thanks also goes out to all the people that are part of the LOFAR-RRL group.

References

Ellingson, S.W., Taylor, G.B., Craig, J., et al.: IEEE Trans. Antennas Propag. **61**, 2540 (2013)
Gordon, M.A., Sorochenko, R.L. (eds.): Radio Recombination Lines. Astrophysics and Space Science Library, vol. 282. Springer, Berlin (2009)
Greisen, E.W.: In: Heck, A. (ed.) Information Handling in Astronomy - Historical Vistas. Astrophysics and Space Science Library, vol. 285, p. 109 (2003)
McMullin, J.P., Waters, B., Schiebel, D., Young, W., Golap, K.: In: Shaw, R.A., Hill, F., Bell, D.J. (eds.) Astronomical Data Analysis Software and Systems XVI. Astronomical Society of the Pacific Conference Series, vol. 376, p. 127 (2007)
Offringa, A.R.: AOFlagger: RFI Software. Astrophysics Source Code Library (2010)
Offringa, A.R., de Bruyn, A.G., Zaroubi, S., et al.: Astron. Astrophys. **549**, A11 (2013)
Oonk, R., Morabito, L., Salgado, F., et al.: Advancing Astrophysics with the Square Kilometre Array (AASKA14), vol. 139 (2015)
Oonk, J.B.R., van Weeren, R.J., Salas, P., et al.: Mon. Not. R. Astron. Soc. **465**, 1066 (2017)
Pandey, V.N., van Zwieten, J.E., de Bruyn, A.G., Nijboer, R.: In: Saikia, D.J., Green, D.A., Gupta, Y., Venturi, T. (eds.) The Low-Frequency Radio Universe. Astronomical Society of the Pacific Conference Series, vol. 407, p. 384 (2009)
Peters, W.M., Lazio, T.J.W., Clarke, T.E., Erickson, W.C., Kassim, N.E.: Astron. Astrophys. **525**, A128 (2011)
Romein, J.: LOFAR ASTRON Technical Report (2008). http://www.astron.nl/~romein/papers/BandPass/bandpass.pdf
Salas, P., Oonk, J.B.R., van Weeren, R.J., et al.: Mon. Not. R. Astron. Soc. **467**, 2274 (2017)
Salgado, F., Morabito, L.K., Oonk, J.B.R., et al.: Astrophys. J. **837**, 142 (2017)
Shaver, P.A.: Pramana **5**, 1 (1975)
Tasse, C., van der Tol, S., van Zwieten, J., van Diepen, G., Bhatnagar, S.: Astron. Astrophys. **553**, A105 (2013)
Tingay, S.J., Goeke, R., Bowman, J.D., et al.: PASA **30**, e007 (2013)
van Haarlem, M.P., Wise, M.W., Gunst, A.W., et al.: Astron. Astrophys. **556**, A2 (2013)
Walmsley, C.M., Watson, W.D.: Astrophys. J. **260**, 317 (1982)

Chapter 13
Particle Physics with LOFAR

Stijn Buitink

Abstract High energy cosmic rays produce cascades of secondary particles in the atmosphere, known as air showers, which emit very short radio flashes (several tens of nanoseconds). The detection of these short radio flashes requires the use of the Transient Buffer Boards (TBBs). In this chapter we will discuss how to use the TBBs, how to calibrate the data, and some typical processing techniques. We round off with an introduction to cosmic ray air shower analysis with LOFAR.

13.1 The Transient Buffer Boards

13.1.1 Description of Buffer Boards

During LOFAR observations, incredible amounts of data are being digitised. The receiver units (RCUs) sample the antenna signals at 160 or 200 MHz with 12 bit A/D converter. Per station, either 96 single-polarised antennas or 48 double-polarised antennas can be processed simultaneously, which amounts to 27 Gb/s. Clearly, the amount of data that are generated in an observation using many stations for several hours is impossible to store. Depending on the observation strategy, the data rate is reduced at station-level by forming beams and selecting subbands. It is however possible to access the raw data before any processing has been done. Directly after digitisation, all data are temporarily stored on ring buffers on dedicated Transient Buffer Boards (TBBs). The TBBs can hold 5 s of data for all active antennas. It is possible to further extend this time window by storing data from fewer antennas or by storing subband data. The signal connections are illustrated in Fig. 13.1.

There are several reasons why it can be interesting to look at the raw time series data. A practical reason is that it can serve as a diagnostic tool; after beam

S. Buitink (✉)
Vrije Universiteit Brussel, Ixelles, Belgium
e-mail: s.buitink@vub.ac.be

G. Heald et al. (eds.), *Low Frequency Radio Astronomy and the LOFAR Observatory*, Astrophysics and Space Science Library 426,
https://doi.org/10.1007/978-3-319-23434-2_13

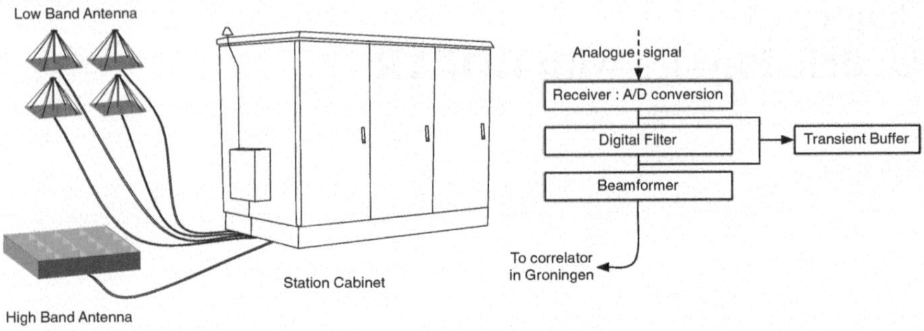

Fig. 13.1 Signal flow at station level. The signals from low-band or high-band antennas are digitised in the station cabinet. All data are temporarily stored on ring buffers, either directly after digitisation (raw timeseries data) or after the digital filter (sub-band data). The data are beam formed before it is send to the central correlator. Credit: van Haarlem et al., A&A, 556, 2, 2013, reproduced with permission © ESO

forming it is no longer possible to identify problems with individual antennas. But there are also great scientific possibilities: the TBBs give LOFAR the power to record very short transient signals. The most advanced application of the TBBs is currently the measurement of radio pulses produced by cosmic-ray air showers in the Earth's atmosphere. These pulses have typical time widths of tens of nanoseconds, and are often confined to only ~10 time samples. It is also possible to search for astronomical transients, e.g. giant pulses from radio pulsars, which are of millisecond time scale. Because the buffered data are not beam formed it is not necessary to know the arrival direction of the signal beforehand. It is, however, necessary to design a triggering algorithm that can make decisions within the 5 s that the data are stored.

13.1.2 Reading Out the TBBs

The TBBs can be read out by sending a dump command. When this happens, the buffers are frozen and read out over the Wide Area Network (WAN) directly to the storage section of the central processor cluster. Because of this freeze, any observation that relies on TBB dumps has to account for dead time during the read-out. However, it is possible to read-out only a (small) fraction of the ring buffer, greatly reducing the dead time and the band width needed for transportation of the data. There are several strategies possible for triggering the TBBs:

- Manual dump. A trigger message can be sent manually to read out the TBB data immediately. This can for example be useful to investigate the noise. Unsuspected noise levels or spikes may indicate that something is wrong with an antenna.
- External trigger. A trigger can be send by an external system. In this case, the trigger message should contain information on what time window needs to be

read out. Cosmic ray air shower detection currently relies on triggers from a particle detector array, LORA (Thoudam et al. 2014), that is constructed at the superterp. When a trigger is issued, a 2 ms window is read out centered around the time where the shower pulse is expected. In principle, it is possible to use any external system for triggering, as long as the trigger message arrives within the 5 s time frame of the ring buffer.

- Local station trigger. At station level, a trigger system can be implemented in the local control unit (LCU). For example, it is possible to trigger on isolated strong pulses with a simple signal-over-threshold trigger. This can be used for cosmic-ray detection, although the false trigger rate is still a problem. Another application is triggering on the very strong pulses of lightning strikes, in order to investigate the electrical processes leading up to a discharge.
- Central trigger. A trigger system can also be implemented at the central processor. This can be used to trigger on weak transient signals that are not visible in individual antennas. An example is the NuMoon project (ter Veen et al. 2015), a search for short radio pulses from neutrinos hitting the Moon. By forming beams on the Moon a high sensitivity to these signals can be achieved. Reading out the raw TBB data allows for a more detailed offline analysis, than would have been possible when only the data present at the central processor are stored.

A wide variety of triggering algorithms can be implemented, tailor-made for specific applications. Note, however, that the development of a new triggering system is quite involved, and can only be done in close collaboration with science support (and probably either the Cosmic Ray KSP or the Transient KSP).

13.2 Calibration

Calibration of time series TBB data relies on different techniques than the calibration of interferometric or beam formed data. The reason for this, is that in $\sim$1 s (or less) of single antenna data, there will not be any astrophysical source visible that has enough signal-to-noise to use it for calibration.

13.2.1 Relative Gain Calibration

The Low Band Antennas (LBAs) measurements are dominated by Galactic sky noise. Since this background is partially polarised the noise levels in the two dipoles of a single antenna are in general different. However, all dipoles with the same orientation detect the same level of noise. This can be used to calibrate the antennas relative to each other.

Since the Galaxy moves across the sky as the Earth rotates, the noise level in each antenna is a function of the Local Sidereal time (LST). Figure 13.2 displays the

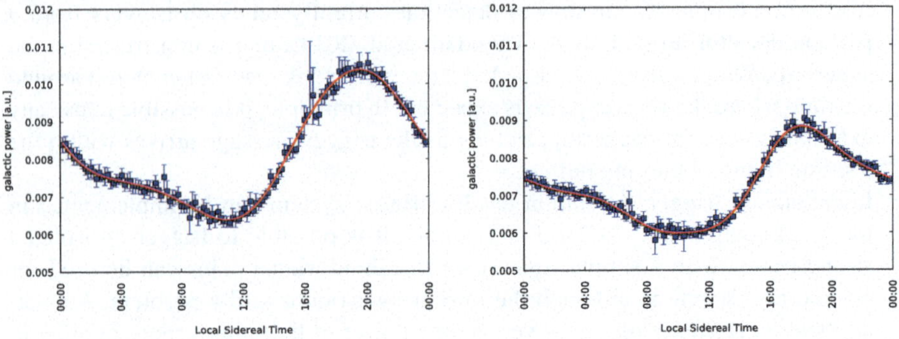

Fig. 13.2 Integrated spectral power normalised to the band width (after RFI removal) as a function of LST for both instrumental polarisations. Left: polarisation X (NE-SW). Right: polarisation Y (NW-SE). Credit: Schellart et al., A&A, 560, 98, 2013, reproduced with permission © ESO

variation in the noise level during the day for the X and Y polarisation of a particular antenna. The data are fitted with an second order Fourier series. The curves that are found in this way are used as a time-dependent reference, i.e. a gain correction is applied to all antennas so that the noise level has the same value across all antennas (but still varies with time).

13.2.2 Absolute Gain Calibration

At the moment, there is no absolute calibration for the TBB data yet, but several approaches are employed to establish one. One possibility is to measure the received power from a known source. Since astrophysical sources are not strong enough, a artificial source has to be used. Two campaigns have been undertaken. First, a reference antenna was attached to an octocopter and flown over the superterp. In summer 2014, more accurate measurements were performed with an emitter that was hung from a wooden crane.

Typical complications with such measurements are that the far-field approximation that is used to calculate the power of the received signal is not entirely precise. Secondly, the reference antenna moves in the wind, which causes variations in the strength and polarisation of the received signal. Finally, the calibration relies on an earlier calibration of the reference antenna itself which was done by the manufacturer but is hard to verify.

An alternative approach is to calibrate on the Galactic background noise. The total noise can be calculated by folding a model of the Galactic radio background with the antenna response model for a single dipole, and integrating over the entire sky.

A combination of the results of the octocopter and crane campaigns, and the Galactic background calculations, is currently in progress.

13.2.3 The Antenna Model

The LBAs have a large field of view and the antenna gain is a complex function of direction, polarisation, and frequency. Therefore, a radio pulse detected in a single antenna cannot be properly calibrated unless its direction is known. Typically, this direction can be found by using the differences in arrival time between pulses in different antennas. Once the direction is known the antenna pattern can be unfolded to calculate the electromagnetic field of the incoming signal.[1]

The antenna pattern for the LBA is simulated with the WIPL-D software package. This program calculates the electric fields generated inside the dipole arms by plane waves from different arrival directions, with different frequencies. The antenna output voltage is then calculated using an equivalent circuit that is a voltage source with an internal resistance equal to the antenna impedance.

The antenna response can be described by the Jones matrix, a complex 2×2 matrix, that translates the field strength of the incoming wave to the output voltages:

$$\begin{pmatrix} V_X \\ V_Y \end{pmatrix} = \begin{pmatrix} J_{X\theta} & J_{X\phi} \\ J_{Y\theta} & J_{Y_\phi} \end{pmatrix} \begin{pmatrix} E_\theta \\ E_\phi \end{pmatrix}, \tag{13.1}$$

where $J_{X\theta}$ is the complex response of the antenna and amplifier of the X-dipole to a wave purely polarized in the $\hat{\mathbf{e}}_\theta$ direction.

The WIPL-D software produces Jones matrices that are calculated for a grid of arrival direction angles and frequencies. The matrices corresponding to directions and frequencies between grid points can be found by interpolation. Now, the electric field strength of the incoming pulse can be found by applying the inverse Jones matrices to measured voltages.

13.3 Working with Time Series Data

While details of data analysis always depend on the particular observation that is being performed, there are several techniques that are universal and will be applicable to any observation that involves the detection of short pulses.

[1] In practice, the unfolding can change the pulse shape and thus the arrival time, which leads to a different arrival direction reconstruction. An iterative approach can than be applied until a self-consistent solution is found.

13.3.1 RFI Cleaning

Raw time series data can contain strong contributions from human-generated radio frequency interference (RFI). RFI can be very strong and be the dominant source of power in the data. Nonetheless, it can be digitally filtered out because it has characteristics that are very different from the short signal pulses. RFI at these frequencies is usually narrow-band, which means it persists over time, but has a very well-defined frequency. Short pulses, on the other hand, have a very broad frequency spectrum and are localised in time. RFI sources can thus be easily identified in the frequency domain, where they appear as narrow spikes.

A frequency spectrum of a time series data set can be produced by splitting the data into blocks, taking the Fast Fourier Transform (FFT) of each block, and then calculate the average of all FFTs. The size of the block sets the resolution of the frequency spectrum. For example, a block of 2^{16} time samples of 5 ns, has a total length of 33 ms, which gives a frequency resolution of 3 kHz, which is enough to resolve most RFI lines.

There are now various strategies to identify and remove the RFI lines. A standard approach is to fit the continuous spectrum with a polynomial function, and flag the frequencies for which the power is far above this baseline fit. A possible problem with this approach is that in the presence of strong RFI it is hard to fit the correct baseline and the technique becomes unstable.

An alternative approach, that is used in cosmic-ray detection, is based on the fact that RFI sources have a stable phase, while noise has a random phase. By imposing a constraint on the maximum allowed phase stability per frequency bin, RFI lines can be flagged without the need to know the correct shape of the baseline of the continuous spectrum.

Figure 13.3 shows an example of LBA data that have been cleaned from RFI.

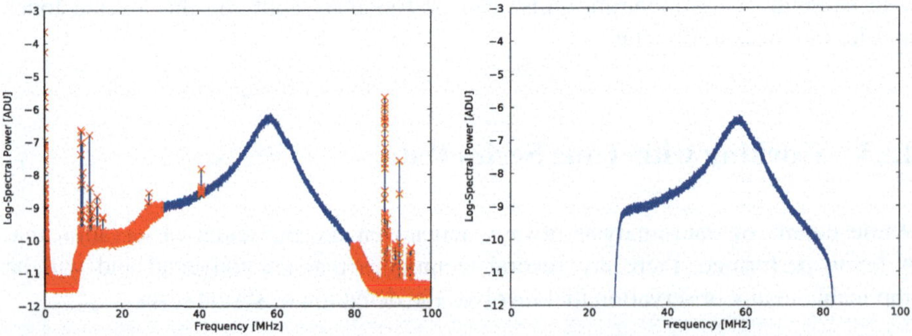

Fig. 13.3 The frequency spectrum of the original data (blue lines in left panel) contains many strong RFI lines. These lines can be identified and flagged (red crosses in left panel) to produce a clean spectrum (right panel). Note that a 30–80 MHz window function was also applied to obtain the final spectrum. Credit: Schellart et al., A&A, 560, 98, 2013, reproduced with permission © ESO

13.3.2 Pulse Finding

After the data have been cleaned from RFI it can be transformed back to the time domain to search for pulses. Usually this is done by a simple over-threshold search, where the whole data set is scanned for amplitudes that are above a predefined threshold. However, depending on the phase, the maximum amplitude can be either up or down. Even worse, when the phase is imaginary ($\pm i$), the signal amplitude is zero in that particular time bin.

A reliable technique is to calculate the Hilbert envelope of the time series data, which is defined as:

$$A(t) = \sqrt{x^2(t) + \hat{x}^2(t)}. \tag{13.2}$$

where $\hat{x}(t)$ is the *Hilbert transform*, or imaginary propagation, of the signal $x(t)$ defined by

$$\mathscr{F}(\hat{x}(t))(\omega) = -i \cdot \mathrm{sgn}(\omega) \cdot \mathscr{F}(x(t))(\omega) \tag{13.3}$$

where $\mathscr{F}$ denotes the Fourier transform. Figure 13.4 shows a cosmic-ray pulse detected by an LBA. The blue curve is the (up sampled) raw signal, and the red

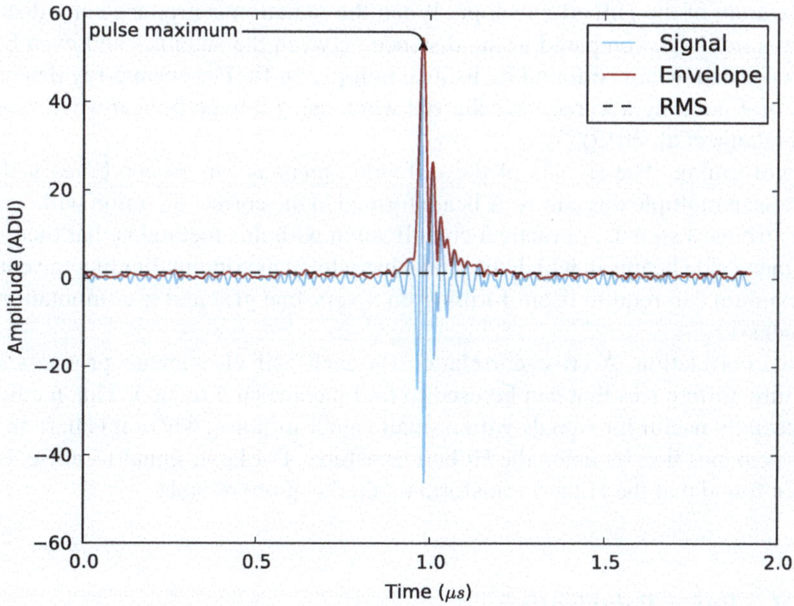

Fig. 13.4 Air shower radio pulse. The raw signal is plotted in blue, and the Hilbert envelope in red. The root mean square of the raw signal is indicated by the dashed line. Credit: Schellart et al., A&A, 560, 98, 2013, reproduced with permission © ESO

curve is the Hilbert envelope. Note that this works on any signal shape and that the envelope can be non-zero even when the original signal is.

13.3.3 Fitting Pulse Arrival Directions

The arrival direction of a pulse can be reconstructed by combining the signals of multiple antennas. Several approaches can be considered:

- Plane wave fit. Given the arrival times t of pulses that were found in the data of antennas at positions (x, y), the best fitting solution for a plane wave

$$ct = Ax + By + C, \tag{13.4}$$

  can be found using a standard least squares approach. From A and B the Cartesian directions ϕ, θ can be extracted as:

$$A = \sin(\theta)\sin(\phi), \tag{13.5}$$

$$B = \sin(\theta)\cos(\phi). \tag{13.6}$$

  A sensible choice for the definition of the arrival direction of the pulse can be the maximum of the Hilbert envelope. When the radiation is produced at a distance that is not large compared to the distance between the antennas and even better reconstruction can be found by using a non-planar fit. For cosmic-ray detection, the best pointing accuracy is achieved when using a hyperbolical wave front fit (Corstanje et al. 2015).
- Beamforming. The signals of the different antennas can be combined to form beams in multiple directions. A beam formed in the correct direction will contain the strongest signal. A practical complication with this method is that there will be many side beams in which the signal has a local maximum. Finding the correct maximum can require beam forming on a very fine grid and is computationally expensive.
- Cross-correlation. A cross-correlation for each pair of antennas provides a set of time differences that can be used to find the arrival direction. This method is especially useful for signals with a small signal-to-noise, where it is hard to find the peak position by using the Hilbert transform. For large signal-to-noise, it has been found that the Hilbert transform method is more reliable.

13.3.4 Pulse Polarization

The polarisation of a pulse can be studied by calculating its Stokes parameters (see Chap. 10). Usually, the best result is obtained by averaging over a reasonable number of bins, n (Schellart et al. 2014):

$$I = \frac{1}{n}\sum_{i=0}^{n-1}(E_{i,x}^2 + \hat{E}_{i,x}^2 + E_{i,y}^2 + \hat{E}_{i,y}^2), \tag{13.7}$$

$$Q = \frac{1}{n}\sum_{i=0}^{n-1}(E_{i,x}^2 + \hat{E}_{i,x}^2 - E_{i,y}^2 - \hat{E}_{i,y}^2), \tag{13.8}$$

$$U = \frac{2}{n}\sum_{i=0}^{n-1}(E_{i,x}E_{i,y} + \hat{E}_{i,x}E_{i,y}), \tag{13.9}$$

$$V = \frac{2}{n}\sum_{i=0}^{n-1}(\hat{E}_{i,x}E_{i,y} - E_{i,x}E_{i,y}). \tag{13.10}$$

where $E_{i,j}$ is sample i of electric field component j and $\hat{E}_{i,j}$ its Hilbert transform. For an elliptically polarized signal one can calculate from the Stokes parameters the angle that the semi-major axis of the polarization ellipse makes with the $\hat{\mathbf{x}}$ axis

$$\psi = \frac{1}{2}\tan^{-1}\left(\frac{U}{Q}\right). \tag{13.11}$$

Additionally the *degree of polarization* is calculated which is defined to be the fraction of the power in the polarized component of the wave

$$p = \frac{\sqrt{Q^2 + U^2 + V^2}}{I}. \tag{13.12}$$

13.4 Introduction to Cosmic Ray Analysis

13.4.1 Radio Emission from Air Showers

Cosmic rays are the most energetic particles in the Universe and are measured at Earth with various techniques over an enormous energy range. There are still many open questions about their origin. It is generally assumed that cosmic rays below 10^{17} eV come from Galactic sources. Supernova remnants are expected to be the main producers of Galactic cosmic rays (Blasi 2013), while the most energetic cosmic rays, up to 10^{20} eV are likely to originate from extragalactic sources, like gamma-ray bursts and active galactic nuclei. At what energy the transition from Galactic to extragalactic origin takes place is still a mystery. Most models place it somewhere between 10^{17} and 10^{19} eV.

Detailed measurements of the cosmic-ray mass distribution can help to disentangle the different components in the cosmic-ray flux. The highest energy cosmic rays from the Galaxy are heavy nuclei, like iron. Because of their large charge they can be accelerated up to higher energies in the shockwaves of supernovae. The extragalactic component, on the other hand, is expected to be dominated by protons and light nuclei, at least up to 10^{19} eV.

There are several techniques to measure cosmic-ray masses. The most accurate techniques are based on measuring the development of air showers, cascades of secondary particles, that are initiated by cosmic rays in the atmosphere (Kampert and Unger 2012). The shower maximum, X_{max} is defined as the atmospheric depth (measured in g/cm^2) at which the number of secondary particles reaches its maximum. Light nuclei penetrate deeper in the atmosphere than heavy nuclei, and therefore the air showers they produce tend to have a larger X_{max}. However, because the development of air showers is a stochastic process, there are large fluctuations in X_{max} for a given cosmic-ray mass and energy. For light particles, this spread is larger than for heavy nuclei.

Air showers can be detected by the radio emission they emit (Falcke et al. 2005). This radiation comes mainly from a transverse current that is induced by the geomagnetic field. Showers contain many electrons and positrons moving at relativistic speeds towards the ground. The Earth's magnetic field accelerates these particles in opposite directions, inducing a current (Kahn and Lerche 1966). The strength of this current increases and decreases as the shower develops. Time-varying currents produce radiation, which is strongly beamed in the forward direction because of the relativistic speeds of the particles. A secondary source of emission comes from the build-up of a negative charge excess in the shower front, from knock-out electrons of atmospheric molecules (Askar'yan 1965).

The radio signal emitted by the shower carries information about the longitudinal development of the shower, and can be used to reconstruct X_{max}, which allows us to study the cosmic-ray mass composition.

13.4.2 Air Shower Reconstruction with LOFAR

The radio emission from cosmic-ray air showers is detected by LOFAR with the TBBs. The LORA particle detector array measures the secondary particles in the air shower as they reach the ground. When enough detectors have a signal above a threshold, a trigger is sent to all core stations, and 2 ms of data are read out from the buffers of all antennas. This observation mode is always running in the background of regular LOFAR observations (piggy-back mode). Hence, which antennas are active and read out depends on the settings of the current observations. For cosmic rays, the LBA data are more useful than HBA data. The HBA tiles are individually beam formed with an analogue beamformer in a preselected direction, which generally does not coincide with the arrival direction of the air shower. Properly calibrating a signal in a side beam is very hard, and only using the air showers that happen to come from the beam direction seriously limits the statistics. For LBA observations, on the other hand, all beam forming is done at station level, i.e. *after* the buffering stage. However, HBA data have been successfully used to derive information on the emission process at higher frequencies (Nelles et al. 2015).

Air shower radio footprints have typical sizes of hundreds of meters in diameter. On the superterp, showers are often detected by multiple (sometimes up to nine) stations. That means that a pulse is found in the data of hundreds of antennas. This allows for a very precise reconstruction of the radiation pattern on the ground. First, the signals from all antennas are cleaned from RFI, using the phase stability method described in Sect. 13.3.1. The antenna response is unfolded using the arrival direction found by the LORA particle array. Then, a pulse search is performed on the cleaned data using the Hilbert envelope procedure described in Sect. 13.3.2. When the pulses are identified, the direction of the pulse is found by performing a plane-wave fit. Since this direction can be different from the original LORA reconstruction, the steps of antenna response unfolding, pulse search, and direction fitting, are repeated until a self-consistent solution is found.

For each pulse, the maximum amplitude and of the Hilbert envelope, the total pulse power, pulse arrival time, and polarisation are calculated. Figure 13.5 shows an event display for a typical shower. The left panel shows the pulse amplitudes and arrival times, and a reconstructed shower core position and direction. The right panel shows the polarisation angle. This angle is a function of observer position relative to the shower core position, and is in good agreement with theoretical prediction (Schellart et al. 2014). In fact, the polarisation can be used to distinguish air showers from short RFI pulses, or even from air shower pulses that were influenced by strong atmospheric electric fields as are present in thunderstorms.

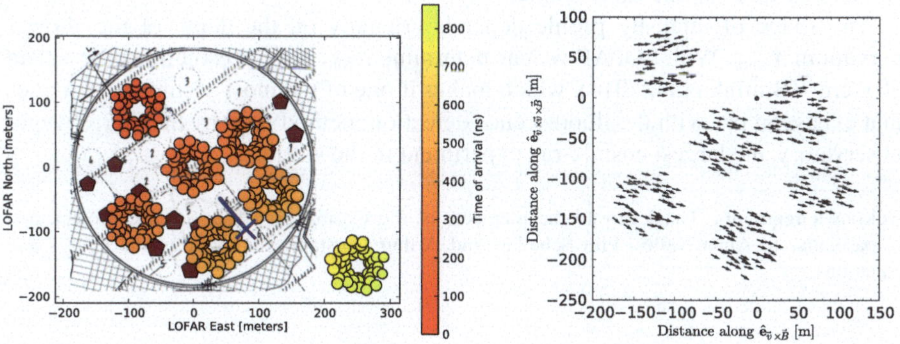

Fig. 13.5 Air shower reconstruction with LOFAR. In the left panel the coloured circles indicate the antennas that detected an air shower radio pulse. Their color indicate the arrival time, while their size reflects the strength of the signal. The LORA particle detectors are indicated by the pentagons. Their size scales with the measured particle density on the ground. The blue "X" marks the reconstructed shower core, and the line emerging from it indicates the reconstructed arrival direction. The right plot displays the polarisation angle that has been measured by each antenna for a different air shower, which has its core at the origin of the plot. Credit: Schellart et al., A&A, 560, 98, 2013, reproduced with permission © ESO

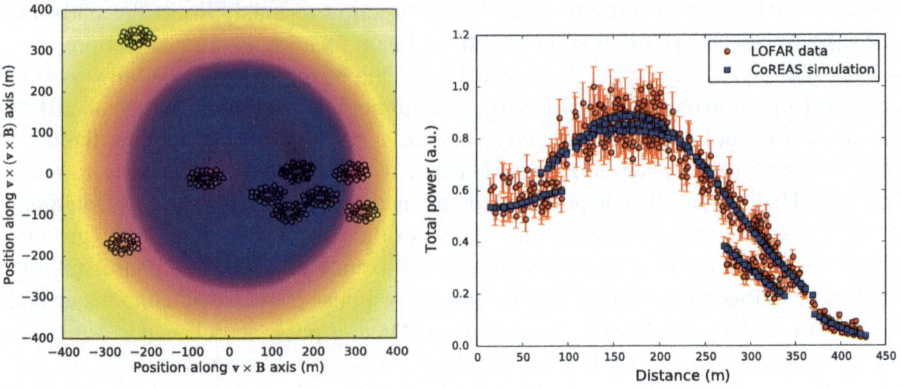

Fig. 13.6 Comparison of data to simulation. The measured power is fitted to a simulated radio profile (left panel). The one-dimensional lateral distribution (right panel) is not a single-valued function of distance to the shower axis

For each detected shower, simulations are produced using state-of-the-art simulation software packages CORSIKA and CoREAS. In these codes, all particles interaction in the air shower are simulated and the radio pulse generated by the collective behaviour of all shower charges is calculated from first principles. From these simulations, we obtain two-dimensional radiation intensity profiles, that have complicated asymmetrical shapes due to the interference between the different emission processes. Figure 13.6 shows a comparison between the measured pulse power in the antennas and a simulated two-dimensional profile. The fit is excellent with a reduced χ^2-value of 1.1. It is clear from the right panel that the measurements cannot be fit with a one-dimensional function.

The shape of intensity profile depends crucially on the depth of the shower maximum $X_{\max}$. With LOFAR we can determine $X_{\max}$ with a resolution of less than 20 g/cm^2 (Buitink et al. 2014), which makes it one of the most accurate techniques that exists, on par with the fluorescence detection method used by the Pierre Auger observatory, the largest cosmic-ray experiment in the world.

Acknowledgements Thanks to all the members of the Cosmic Ray Key Science Project, and in particular to Anna Nelles, Pim Schellart and Arthur Corstanje for many of the figures and equations.

References

Askar'yan, G.A.: Sov. J. Exp. Theor. Phys. **21**, 658 (1965)
Blasi, P.: Astron. Astrophys. Rev. **21**, 70 (2013)
Buitink, S., Corstanje, A., Enriquez, J.E., et al.: Phys. Rev. D **90**, 082003 (2014)
Corstanje, A., Schellart, P., Nelles, A., et al.: Astropart. Phys. **61**, 22 (2015)
Falcke, H., Apel, W.D., Badea, A.F., et al.: Nature **435**, 313 (2005)

Kahn, F.D., Lerche, I.: R. Soc. Lond. Proc. Ser. A **289**, 206 (1966)
Kampert, K.-H., Unger, M.: Astropart. Phys. **35**, 660 (2012)
Nelles, A., Schellart, P., Buitink, S., et al.: Astropart. Phys. **65**, 11 (2015)
Schellart, P., Buitink, S., Corstanje, A., et al.: J. Cosmol. Astropart. Phys. **10**, 14 (2014)
Schellart, P., Nelles, A., Buitink, S., et al.: Astron. Astrophys. **560**, A98 (2013)
ter Veen, S., Buitink, S., Corstanje, A., et al.: In: Borisov, A.S., Denisova, V.G., Guseva, Z.M., Kanevskaya, E.A., Kogan, M.G., Morozov, A.E., Puchkov, V.S., Pyatovsky, S.E., Shoziyoev, G.P., Smirnova, M.D., Vargasov, A.V., Galkin, V.I., Nazarov, S.I., Mukhamedshin, R.A. (eds.) 34th International Cosmic Ray Conference (ICRC2015), vol. 34, p. 684 (2015)
Thoudam, S., Buitink, S., Corstanje, A., et al.: Nucl. Instr. Meth. Phys. Res. A **767**, 339 (2014)
van Haarlem, M.P., Wise, M.W., Gunst, A.W., et al.: Astron. Astrophys. **556**, A2 (2013)

Chapter 14
High Time Resolution with LOFAR

Jason Hessels and Richard Fallows

Abstract LOFAR achieves high time resolution (<1 s samples) using its various beam-formed modes. These modes are vital for observations of pulsars, fast radio transients, the Sun, (exo)planets, flare stars and dynamic spectra of scintillating sources—to name a few examples. Here we describe the observing modes and tools available to process raw LOFAR beam-formed data, including the standard pulsar pipeline ('PulP'; `pulp.py`) and further analysis—with a focus on pulsar dedispersion, folding, RFI excision, polarimetry and searching. We also describe a general purpose dynamic spectrum toolkit.

14.1 Scope of This Chapter

This chapter discusses how one can undertake high-time-resolution radio astronomy with LOFAR, using its standard 'beam-formed' observing modes. We also include some novel examples of how station beam forming can be more flexibly configured than in the standard observing modes. The analysis of pulsar data and associated issues is described in detail; a brief discussion on the analysis of dynamic spectra is also included.

J. Hessels (✉)
ASTRON, The Netherlands Institute for Radio Astronomy, Dwingeloo, The Netherlands

Anton Pannekoek Institute for Astronomy, University of Amsterdam, Amsterdam, The Netherlands
e-mail: hessels@astron.nl

R. Fallows
ASTRON, The Netherlands Institute for Radio Astronomy, Dwingeloo, The Netherlands
e-mail: fallows@astron.nl

G. Heald et al. (eds.), *Low Frequency Radio Astronomy and the LOFAR Observatory*, Astrophysics and Space Science Library 426,
https://doi.org/10.1007/978-3-319-23434-2_14

14.2 Motivation for High Time Resolution with LOFAR

The majority of LOFAR astronomers do their science using the instrument's imaging mode (hereafter IM mode), which is described in detail in this reference book, as well as van Haarlem et al. (2013). While IM mode is obviously the right choice for any application that requires reasonably good angular resolution and/or imaging capabilities, its main limitation is that the integration time used to produce the visibilities is typically at least 1 s or longer for LOFAR. For low-frequency radio science that requires higher *time* resolution, or for which imaging is irrelevant, LOFAR's beam-formed modes (hereafter BF modes) are needed.[1] In LOFAR observing Cycles to date, BF modes were requested for roughly a third of the granted observing time. As such, next to normal imaging they form a significant part of the telescope's scientific program. The LOFAR BF modes are described in Stappers et al. (2011) as well as van Haarlem et al. (2013), and further practical details are given in this chapter.

LOFAR's 200-MHz station clocks are typically used for sampling the raw signal[2], and each station can provide up to 244/488 subbands in 16-bit/8-bit mode respectively. The resulting 0.1953125-MHz subbands have a sampling rate of 5.12 μs, which in most cases is the maximum achievable time resolution that the system provides. By inverting the first polyphase filter (1PPF; this is done at LOFAR station level), it is possible to recover an even higher time resolution, but this is not done routinely. Likewise, capturing the raw antenna signals from the Transient Buffer Boards (TBBs) affords nanosecond time resolution, though only a few seconds of data can be captured at any given time before filling the TBB memory. These non-standard high-time-resolution modes are described in more detail in Chap. 13 on LOFAR cosmic-ray observations; they certainly offer advantages for other types of high-time-resolution radio astronomy as well. Thus, in most cases, 5.12 μs is the base time resolution available to the user, and it can be degraded in favor of higher frequency resolution (see Sect. 14.3).

There is a growing variety of astronomical sources that show variations on sub-second timescales. Radio pulsars are a classical example, as are solar bursts and planetary emission. A growing variety of radio pulsars have been identified since their original discovery almost 50 years ago (Hewish et al. 1968). This includes millisecond pulsars (Backer et al. 1982, which rotate hundreds of times more rapidly than 'normal' pulsars), pulsars that switch between two stable modes of radio and high-energy emission (Hermsen et al. 2013), radio pulsars that pulse sporadically (so-called 'rotating radio transients', RRATs; McLaughlin et al. 2006), are only intermittently active (Kramer et al. 2006), have quasi-cyclic changes in

[1] Nonetheless, note that fast (millisecond) imaging is a growing area of technical development.

[2] A 160-MHz clock is also available, but it is not (as yet) routinely used. The 160-MHz clock provides 0.15625-MHz subbands and 6.4 μs sampling.

their profile shape (Lyne et al. 2010), or that switch back-and-forth to a low-mass X-ray binary state (Papitto et al. 2013; Stappers et al. 2014; Bassa et al. 2014). Recent pulsar surveys have also discovered a phenomenon called 'fast radio bursts' (Lorimer et al. 2007; Thornton et al. 2013; Spitler et al. 2014), which appear to represent a population of radio transients at cosmological distances. These discoveries have spurred intense interest in further exploring the transient radio sky at high time resolution using sensitive radio telescopes with large fields-of-view (FoVs).

Thus far with LOFAR, pulsar studies have probed the properties of the interstellar medium, including scintillation (Archibald et al. 2014) and dispersion (Hassall et al. 2012). Some of the well-known moding pulsars like PSRs B0943+10 and B0823+26 have been studied in great detail and new phenomena have been discovered (Sobey et al. 2015; Hermsen et al. 2013; Bilous et al. 2014). Millisecond pulsars, which are particularly challenging to detect in the LOFAR band, have also been studied (Dolch et al. 2014; Stovall et al. 2014; Kondratiev et al. 2016). Using LOFAR's large fractional bandwidth, the evolution of the pulse profile with frequency has been studied in detail (Hassall et al. 2012, 2013; Pilia et al. 2016). Likewise, pulsar polarimetry has provided insights into the pulsar emission mechanism (Noutsos et al. 2015), along with unprecedented accuracy in Faraday rotation measurements, which provide information about the intervening magnetic field in, e.g., the ISM and ionosphere (Sotomayor-Beltran et al. 2013). Finally, pulsar and 'fast transient' surveys are ramping up with LOFAR (Coenen et al. 2014, see also http://www.astron.nl/lotaas/ for a list of LOFAR's first 74 pulsar discoveries), and the wide-field, multi-beaming capabilities of the telescope have also been used to localize and characterize rotating radio transients discovered in other ongoing surveys as well (Karako-Argaman et al. 2015).

LOFAR BF modes also have many uses outside of pulsar science: e.g., Morosan et al. (2014) used 169 tied-array beams to mosaic the Sun and corona and 'image' Type III radio bursts with a higher time cadence than is possible via LOFAR's standard IM mode. Scintillation of compact radio sources due to the interplanetary medium can be used to study the solar wind (Fallows et al. 2013) and pulsar scintillation techniques originally used to study the interstellar medium have been applied to the ionosphere (Fallows et al. 2014). Observations of lightning emission from Saturn and low-frequency emission from Jupiter have also been carried out. In addition, much of this science can be carried out using single stations; novel observing techniques using LOFAR single-station hardware and observations of ionospheric absorption are detailed in McKay-Bukowski et al. (2014).

If you are new to the field, hopefully this chapter will give you the basic knowledge to write your own LOFAR paper based on beam-formed data.

14.3 Observing at High Time Resolution with LOFAR

14.3.1 Brief Summary of Beam Definitions

It is first useful to give a brief overview of the definitions of the various 'beams' which can be formed at various stages in the LOFAR signal chain, as some terminology may be unfamiliar to anyone who has not used LOFAR before. LOFAR beam-forming is hierarchical in nature—forming first beams at HBA tile level, then station level, and then array level—so the picture can be confusing if one is not completely clear on what 'beam' is meant in a particular situation. See also Chap. 2.

- High-band array (HBA) 'tile beam':
  - Each HBA tile contains an analogue beam-former which is used to 'point' the tile (which contains 16 dual-polarization antennas) in a given direction.
  - The half-power beam width is approximately 15–30° over the frequency range 240–110 MHz.
  - Only one beam can be formed at a time.
- Station beam or 'sub-array pointing' (SAP):
  - A digital beam-former at each station is used to combine low-band antennas or HBA tiles within a station to 'point' in a given direction.
  - Capable of forming up to 488 'beamlets' (in 8-bit mode). Each beamlet represents a single pointing direction with a bandwidth of one sub-band (0.1953125 MHz when the 200 MHz sampling clock is used).
  - A SAP is formed from the combination of any number of these beamlets.
  - Multiple SAPs can be formed so long as the total number of beamlets does not exceed the maximum. Thus the bandwidth available to each beam is reduced according to the number of station beams formed.
  - While 488 SAPs (maximum) are possible, more typically only a few are created, with 10s of MHz bandwidth each.
- Array beams formed by the main correlator (COBALT):
  - 'Tied-array beam (TAB)', or 'Coherent Stokes beam (CS)':

    Each SAP returned from multiple stations can be combined coherently to form multiple tied-array beams, each with a different pointing direction.
    These represent the narrowest beams which can be formed with LOFAR, with half-power beamwidths of less than 0.1° depending on frequency and the baselines between the stations that are used.
    Up to approximately 200 beams can be formed, each with the same bandwidth as the SAP(s) used.
    Restricted to using Core stations because only these stations share the LOFAR 'Single Clock' system.

Across the 2-km-wide LOFAR Core, the ionosphere can cause the TABs to mispoint in a dynamic way. This is far less of an issue when using only the 300-m-wide Superterp.

– 'Incoherent beam', or 'Incoherent Stokes beam (IS)':

  Each SAP returned from multiple stations can be combined incoherently (the signals are 'detected' before summing with the appropriate geometrical delay) to form a single incoherent beam for each SAP.
  Any set of LOFAR stations can be used as input.
  Half power beam width wider (i.e. same size as SAP) than for a coherent tied-array beam.
  Experience shows that these beams are more susceptible to the different RFI environments of the individual stations.

– 'Fly's Eye' beam:

  The data from each station are processed by the correlator to average in time and increase frequency resolution as required, but are recorded separately without additional beam-forming.

These definitions and restrictions need to be taken into account when designing a LOFAR observation. For example, a coherent TAB will only return useful data if its pointing direction falls within the SAP, and an HBA SAP will only return useful data if it falls within the HBA tile beam. The following sections describe these in more detail and how they may be used in observations.

14.3.2 Standard Beam-Formed Modes

Each LOFAR BF sub-mode has its own strengths and weaknesses. The currently available suite of modes is limited largely by system management and configuration practicalities, and additional modes may become available in the future (see Sect. 14.7). In general, when we talk about a LOFAR BF mode, we mean an observing mode that creates a combined (i.e. summed) data stream where the fundamental building block is a station beam (SAP). LOFAR stations can in principle point each 'beamlet' (frequency subband) in an independent direction on the sky. In practice, normally only a few station beams are created, in order to provide sufficient bandwidth per beam. LOFAR BF modes thus combine SAPs in different ways (Fig. 14.1). We now describe the main modes.

Coherent Stokes (CS) Mode When SAPs are combined before detection, coherently (in-phase), we refer to this mode as 'Coherent Stokes' (hereafter CS). Perhaps confusingly, CS beams are often also referred to as 'tied-array' beams or 'pencil' beams. The terms are basically interchangeable, but here we will stick to using 'CS beams' exclusively. In principle, CS mode increases the sensitivity linearly with the number of stations added. In practice, there is a loss of sensitivity compared

Fig. 14.1 Illustration of the individual tile beam (left), station SAPs (middle), and multiple CS beams (right). The bar in the bottom-left corner indicates how the picture is zooming out. Credit: Stappers et al., A&A, 530, 80, 2011, reproduced with permission © ESO

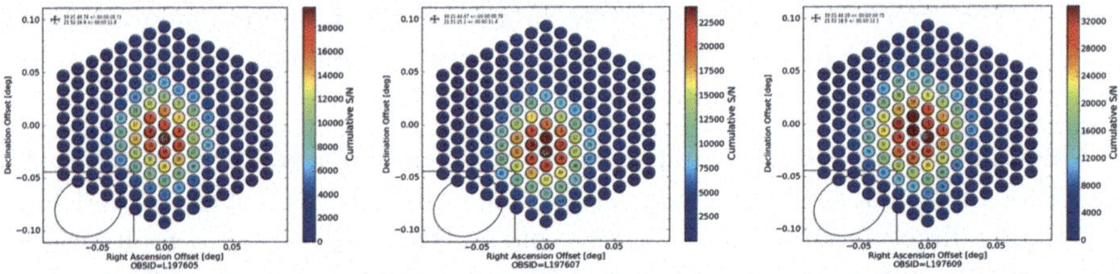

Fig. 14.2 Three consecutive, 2-min LOFAR HBA BF observations (L197605, L197607, and L197609). The time between observations is 20 min. In these observations, 169 CS beams have been closely packed using the 'rings' specification method. At the center of the field is the very bright ($S_{150} \sim 1$ Jy) pulsar B1919+21 (the first pulsar ever discovered). Because the CS beam centers are much closer than their FWHM (~5 arcminutes for these Full Core observations), the pulsar signal appears in many beams. Each of the plotted circles marks a CS beam center (but *not* its size on the sky), and the color indicates the S/N of the pulsar during the observation. As such, they basically trace the CS beam sensitivity on the sky. For comparison, the theoretically expected beam shape, projected onto the sky, is shown in the bottom-left corner. One can see that the centroid of the beam shifts between the observations. This is attributed to differential ionospheric phase delays, which are not calibrated in real time. A movie of a longer sequence of these observations is available here: *http://www.astron.nl/dailyimage/index.html?main.php?date=20140123*. These observations were taken during a period of apparently enhanced ionospheric activity, and it is possible that they are not representative of the average beam-stability conditions. *Figure credit: Wilfred Frieswijk*

with the ideal case because of calibration inaccuracies and ionospheric beam jitter (Fig. 14.2). It is also possible that not all stations are equally sensitive, because of hardware or calibration problems at station level. For a given set of stations, with certain maximum baseline and *uv* distribution, one can think of the CS beam shape and size being approximately equal to the PSF of the synthesized interferometric beam. While a single LOFAR core HBA sub-station has a beam-width (here for simplicity just the FWHM of the main lobe) of $\sim 6°$, the CS beams from the 300-m-wide Superterp (innermost 12 HBA sub-stations) are only $\sim 0.5°$, and the Full

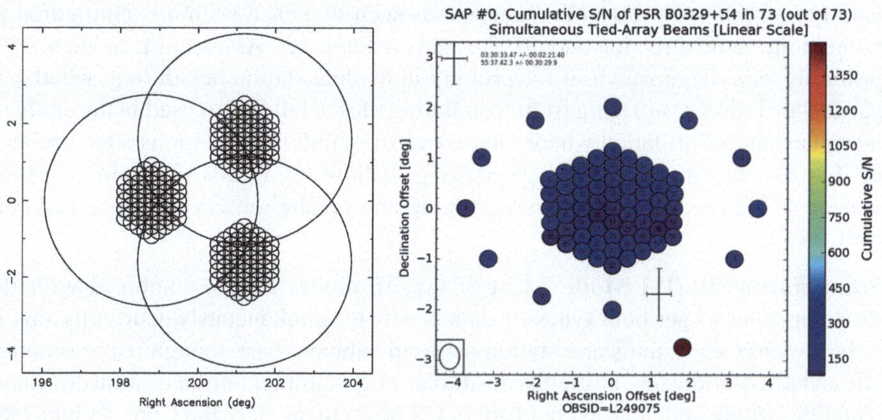

Fig. 14.3 *Left:* Example of how multiple SAPs and CS/IS beams can be used for a survey. Illustrated is the strategy of the LOTAAS survey (http://www.astron.nl/lotaas/), which uses 3 SAPs, each with 1 IS beam, 61 CS beams specified by rings, and 12 CS beams manually specified (not shown here) to point at known sources within the SAP FoV. *Right:* Diagnostic plot of LOTAAS observation L249075. Only SAP number 0 is shown. There is a hexagonal grid of 61 CS beams, and 12 manually pointed CS beams. One of those manually pointed beams is directed to PSR B0329+54, the brightest pulsar in the northern hemisphere. The color scale indicates the S/N of the pulsar in each beam. The theoretical beam shape is shown in the bottom-left corner. *Figure credit: Vlad Kondratiev*

Core (which is ∼2 km across) CS beams are only ∼0.08° (5 arcminutes) wide. For observations of individual known sources, these restricted FoVs are sufficient, but for surveys, multiple CS beams must be generated to recover a larger fraction of the station FoV (Fig. 14.3). It is also worth noting that CS beams can be formed within each SAP. These can be manually specified to point in particular directions on a beam-per-beam basis, or a hexagonal grid of beams can be generated by simply indicating the desired number of hexagonal 'rings' and their radial separation (Fig. 14.3).

Incoherent Stokes (IS) Mode LOFAR's 'Incoherent Stokes' (hereafter IS) mode combines SAPs after detection, i.e. incoherently. This approach maintains the full FoV of the individual stations, regardless of the baseline between them, but increases the sensitivity only as the square-root of the number of stations being combined. In practice, there is an additional hit in sensitivity because IS mode is often less robust to radio frequency interference (RFI). While CS mode often 'correlates out' RFI that is uncommon to the various stations, IS mode data can simply be the sum of the RFI environments at each station. Online excision of RFI on a *per station* basis can help significantly, but this is still being developed (see Sect. 14.7).

Fly's Eye (FE) Mode In some cases, it is desirable to record the station beam signals independently. This mode is called 'Fly's Eye' (hereafter FE). Currently it is only possible to specify FE observations such that each station is configured to point in the same direction and use the same antenna set. As such, FE mode is used primarily as a diagnostic tool for probing individual station health (e.g. whether a particular station is suffering from 'oscillating tiles'), but is also used in the study of interplanetary scintillation where time series from individual stations are correlated off-line to estimate solar wind speeds (e.g. Fallows et al. 2013). A more flexible version of FE mode providing sub-arraying and parallel observing is planned (see Sect. 14.7).

Simultaneous BF/IM Mode Most of the BF modes can be combined with IM mode in order to get both types of data products simultaneously. Currently this is only possible using the same station set (and subband list), though using separate BF and IM station sets is highly desirable. For example, one may want to image using the whole Dutch array, but form a CS beam using only the Core stations (see Sect. 14.7).

14.3.3 Customizing the Beam-Formed Modes

The CS, IS, and FE monikers refer to how SAPs are combined (or not) early in the signal processing chain on COBALT[3] (see other chapters of this book and Mol and Romein (2011), for a detailed description of the processing on the LOFAR beam-former/correlator). This is only the first step in COBALT's online processing, however, next the observation specific parameters are implemented, as we now discuss.

Choice of Time/Frequency Resolution Each of the LOFAR BF modes offers the ability to tune the output time and frequency resolution. Starting from 195/156-kHz subbands, one can further channelize to 16, 32, 64, 128, 256, or more channels per subband using the second polyphase filter (2PPF; this is applied on COBALT). Naturally, this automatically degrades the output time resolution according to the Nyquist limit. It is possible to also further downsample in time or collapse frequency channels in order to throttle the data rate. In practice, the data rate to each CEP2 node should be lower than 700 Mb/s. In some cases, this may necessitate splitting individual beams in frequency across multiple nodes. All of these detailed settings can be changed in MoM and the Scheduler. As the system upgrades to new data storage clusters (e.g. CEP4) the exact data rate limits per node will also likely change.

Choice of Stokes (or Not!) Parameters As a last step in the online BF processing, one must chose what Stokes parameters to save, or whether to circumvent forming these entirely. Recording only Stokes I obviously provides the lowest possible data

[3]Rest in peace sweet, princely BG/P.

rate but, if the science case requires it, Stokes IQUV can also be recorded. For more advanced signal processing applications like coherent dedispersion or cyclic spectroscopy, it is also possible to record the Nyquist-sampled raw voltages. This is also referred to as 'complex voltage' (CV) mode. CV mode data produce roughly 6 Gb/s of data per beam (for 80-MHz bandwidth), and hence such beams are often split into 20 frequency subbands, written to separate files on separate CEP2 nodes.

14.3.4 Other Observing Modes

Standalone Mode Most of the LOFAR international stations in France, Sweden, Germany, the United Kingdom, Poland and Ireland are equipped with on-site backends capable of recording and reducing individual station data (e.g., Serylak et al. 2013). These operate when these stations are not in International LOFAR Telescope (ILT) mode, and might also regularly piggy-back on ILT observations in the future. The standalone backends provide much of the same functionality of COBALT in terms of further channelizing the subbands and tuning the output time resolution. In some cases they also offer real-time scientific data processing, such as searching for individual dispersed radio pulses or folding/dedispersing the data using the known rotational ephemeris of a pulsar.

Raw UDP Dumps It is also possible to dump the stations' raw UDP packets to disk on CEP4. This is not a standardly available mode, and is used primarily for testing COBALT because the raw station data can be 're-played' through COBALT repeatedly in order to test many different observing configurations with the same input data.

14.3.5 Observing Challenges

Propagation Effects Astronomical radio signals are affected by their propagation through the ionized and magnetized inter-stellar medium (ISM).[4] For short-duration signals, these propagation effects can very significantly change the observed signal (compared with the intrinsic signal at the location of emission), *and* they can also greatly degrade our ability to detect such signals in the first place. The relevant propagation effects include (i) scintillation, which is a constructive/destructive interference of the signal with itself due to diffraction and/or refraction (Fig. 14.4, left); (ii) scattering, which causes multi-path propagation of the signal (Fig. 14.4, right); (iii) dispersion, which is a frequency-dependent light travel time in the medium (Fig. 14.5, left); and (iv) Faraday rotation, which rotates the angle of linear

[4]Fast radio bursts are also affected by propagation in the inter-galactic medium, assuming they are of extragalactic origin.

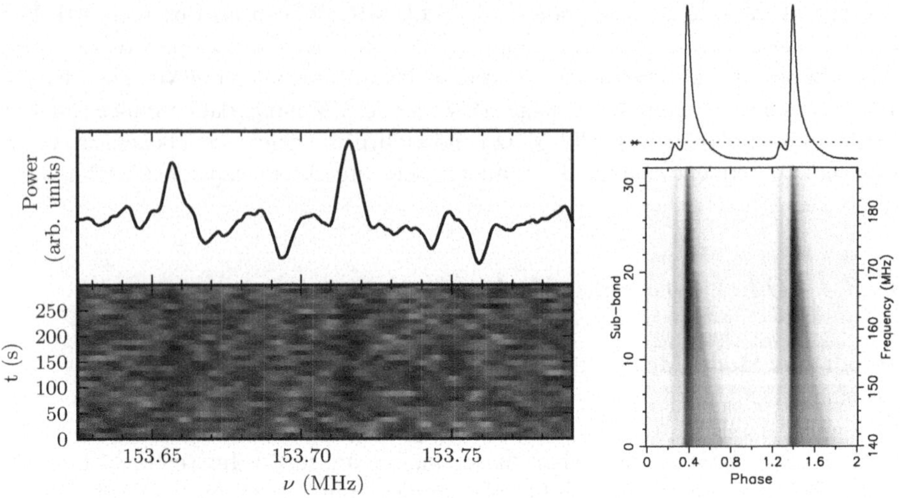

Fig. 14.4 *Left:* A high-resolution dynamic spectrum of millisecond pulsar J1810+1744 (observation L203594; from Archibald et al. ApJ (2014) 790, 22, reproduced by permission of the AAS). Note that only ∼ 160 kHz (less than a single subband) is shown along the *x*-axis, even though 400 subbands (80 MHz total bandwidth) were recorded in total. The spectrum is zoomed to show two faint scintillation features, each only ∼ 2 kHz wide. In order to produce this image of PSR J1810+1744's scintillation structure, cyclic spectroscopy was required to get both the required pulse-phase resolution (to separate this 1.7-ms pulsar's on and off-pulse phase windows) and frequency resolution (to resolve the extremely narrow scintles). *Right:* The pulse profile of PSR B2111+46 becomes progressively more scattered across the LOFAR HBA band (observation L62446). Simply described, the intrinsically narrow main pulse is convolved with a one sided exponential. The summed profile across the band is show at the top, and the profile is repeated over two rotational cycles for clarity

polarization as the signal passes through a magnetized region (Fig. 14.5, right; Fig. 14.6; see also Chap. 10). See Rickett (1990) for an excellent overview of the subject.

Important for LOFAR observations is that these effects are all strongly chromatic, and increase exponentially towards lower frequency—e.g. dispersive delay is proportional to ν_{obs}^{-2} and scattering delay is proportional to roughly ν_{obs}^{-4} (here the exponent will depend on the turbulent properties of the intervening material). This means that signals that are lightly affected by scattering at the top of the LOFAR high-band (110 − 190 MHz) can be very strongly affected at the bottom of the band (Fig. 14.4, right). Correcting for dispersive delay is done routinely through a technique called 'dedispersion'. For incoherent dedispersion, the frequency channels are simply shifted in time in order to compensate for the delay across the band. This is computationally efficient, but leaves residual dispersive smearing within the frequency channels. Coherent dedispersion operates on the raw voltage signals and completely corrects for dispersion. It is more computationally expensive and hence typically only used for individual sources with known dispersion measure (DM) or limited searches in DM space. See Lorimer and Kramer (2004) for more

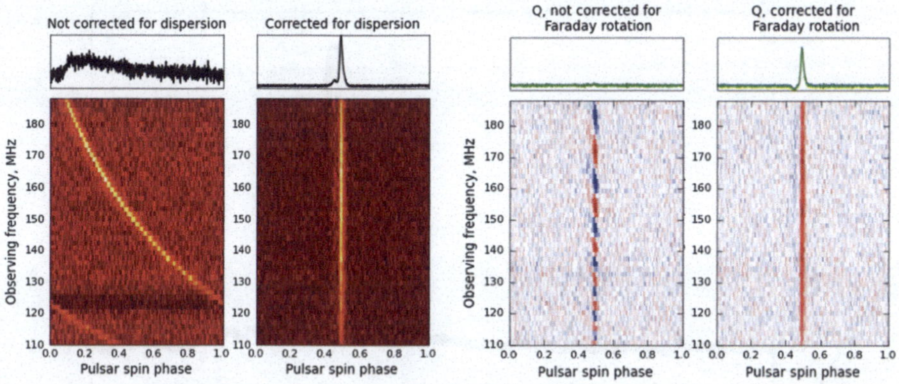

Fig. 14.5 *Left:* Dispersion causes an impulsive radio signal to arrive later towards lower observing frequencies. In the left panel, this effect can be seen for PSR B2021+51 in LOFAR observation L227667, where an incorrect DM of 19.50 pc cm^{-3} has been applied to the data. This DM is off by only $\sim$3 pc cm^{-3} compared with the correct value. In the right panel, the correct DM of 22.5497 ± 0.0006 pc cm^{-3} is applied (this DM is very precisely measured in this single LOFAR observation). In both cases, the top sub-panel shows the integrated pulse profile across the whole observing band. The pulsar is only well detected when the data are properly dedispersed. *Right:* The polarimetric properties of PSR B2021+51 in the same observation are illustrated. The left panel shows Stokes Q oscillating between negative (blue) and positive (red) values as a function of frequency because of Faraday rotation. This happens more rapidly towards lower frequencies. The right panel plots Q after correction for Faraday rotation. In both cases, the top sub-panel shows the Stokes Q component of the pulse profile integrated across the whole observing band. *Figure credit: Anya Bilous*

details. Scattering, which will (roughly speaking) convolve an impulsive signal with a one-sided exponential tail cannot be corrected for.[5]

Ionospheric Effects Coherent sums of the LOFAR station beam signals are only effective for combining the Core stations (the 24 'CSNNN' stations). One reason is because only these stations share the LOFAR Single Clock signal, meaning that their local time standards do not drift with respect to each other at the $5-20$ ns level (as other stations do), which would otherwise render their coherent sum useless. With these stations on a common clock, careful calibration tests have been done to calculate the residual cable delays between stations, which are then removed on a per-polarization basis as a first processing step on COBALT. This works reasonably well, but does not provide a perfect level of coherency. Another issue is that differential phase delays between stations, caused by the ionosphere, become problematic for increasing baselines. While this is likely a negligible problem for beams formed with the Superterp, where the maximum baseline is $\sim$300 m, under certain conditions the effect is large enough to displace a CS beam formed from

[5]Though using, e.g., cyclic spectroscopy, it might be possible to coherently de-scatter if very high S/N data is available.

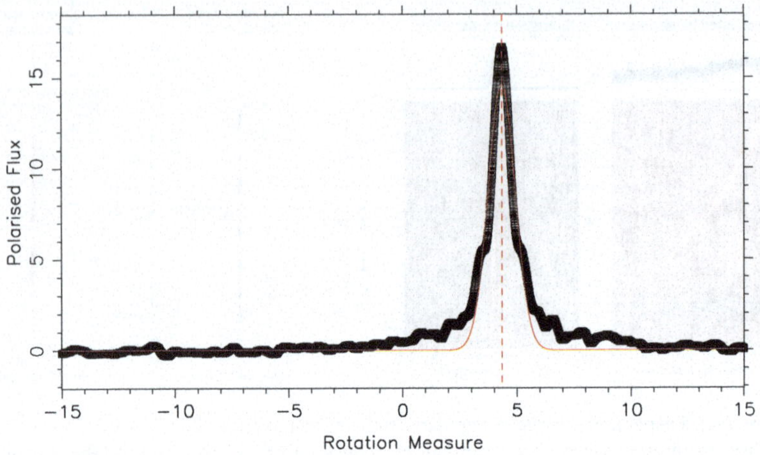

Fig. 14.6 Linearly polarised flux (arbitrary units) as a function of trial RM (rad m^{-2}) using the data shown in Fig. 14.5, right. The location of the peak of the function (shown by the red dashed line) represents the measured RM value towards PSR B2021+51. Note that a precise value can be obtained because the full width half maximum of the peak becomes very narrow ($\sim$1 rad m^{-2} in this case) using lower-frequency data. *Figure credit: Charlotte Sobey*

the Full Core by a significant fraction of its FWHM (5 arcminutes at 150 MHz), see Fig. 14.2. In the absence of a real-time calibration pipeline, the only resort is to use multiple tied-array beams that sample the possible area over which the beam might wobble, thereby ensuring that the source is always near a beam center.

Data Volume Each individual LOFAR station sends $\sim$3 Gb/s of data to COBALT, either as 16-bit or 8-bit samples. It is important to note that COBALT currently only writes output samples as 32-bit floats, meaning a factor 4 expansion in data volume. The highest routinely used output data rate from COBALT to CEP2 is $\sim$40 Gb/s, which is used, e.g., in the LOFAR pulsar surveys (Coenen et al. 2014). Recording 17 TB of raw data per hour is a burden to process and to archive. In the case where raw data must be saved, the Pulsar Pipeline (PulP) can be used to convert the samples to 8-bits. Many science applications also save only data products derived from the raw data. As a future extension, COBALT may also be able to output 8-bit, 4-bit, or even 2-bit data (see Sect. 14.7).

14.3.6 Planning an Observation

Here we discuss some practical considerations for observation planning.

Choosing the Right Stations CS and IS modes provide the sum of multiple station beam signals, and hence are highly sensitive to malfunctioning stations. Special care should be taken to avoid the inclusion of stations that are misbehaving, e.g.

because they contain several HBA tiles with the known 'oscillating tile' problem. LOFAR Science Support performs regular system health observations to identify such problems, and it is usually known which stations should be avoided on a daily basis. One way in which these stations are identified is through the FE mode, by pointing at a bright known pulsar and comparing the resulting signal across stations. The core station CS013 is almost always left out of HBA observations because its HBA X and Y dipoles are rotated by 45° with respect to the other stations.

Choosing CS vs. IS Mode Most science cases will benefit more from using CS mode because of its higher sensitivity and greater robustness to RFI. Nonetheless, some science cases also benefit from the wider FoV of the IS mode. It is worth noting that CS and IS mode can easily be run in parallel, even if many CS beams are formed. One application of such an observation is that the IS beam can serve as a control to separate RFI signals from genuine astrophysical events. The output time and frequency resolutions of the CS and IS data are also independently configurable.

Choosing Time/Frequency Resolution The required output time/frequency resolution is dictated by the science requirements and properties of the astronomical signal being observed. Quite often the required frequency resolution is driven by the need to correct for interstellar dispersion. For example, for low-DM pulsars, one typically uses 16 channels per subband (12-kHz channels), which provides 81.92 µs native time resolution (for the 200-MHz clock). Often this is downsampled to 327.68 µs to reduce the data rate. Higher frequency resolution is necessary for higher-DM pulsars (or low-DM pulsars observed in the low-band), and very high frequency resolution (say 1024 channels per subband) is desirable for spectral line work, where the requirements on time resolution are greatly relaxed.

Choosing the Number of Beams Depending on the experiment, using two CS beams, one on source, and one off source may be desirable. It should be ensured that the 'off' beam is many beam FWHM away from the 'on' beam, but not on the edge of the SAP FoV. As mentioned above, IS beams are sometimes also useful in this regard. The maximum number of CS beams possible is limited by the maximum 40 Gb/s output rate, but also in certain situations by online computing power. Obviously, the desired time/frequency resolution of individual beams has a strong impact on how many can be recorded. For the typically used time/frequency resolution, a maximum of a couple hundred CS beams can be recorded. When using CV mode, only ~ 6 beams can be recorded if 400 subbands are used.

14.4 Analyzing LOFAR Beam-Formed Data

On a *per-beam basis*, analyzing LOFAR beam-formed data for pulsar science is not radically different than the standard procedures used for analyzing single-dish pulsar data. For a detailed introduction to such techniques, see the *Handbook of pulsar astronomy* (Lorimer and Kramer 2004). As such, here we will simply emphasize important LOFAR-specific points. Basic dynamic spectrum analysis is described in Sect. 14.6.

HDF5 Format LOFAR BF data is written in Hierarchical Data Format 5 (HDF5 format), using a header structure which is described in detail in LOFAR Interface Control Document 3 *Beam-formed data* (ICD3; see also Alexov et al. (2010) and Alexov et al. (2012)). The LOFAR Data Access Library (DAL) can be used to work with these files at a low level. Typically, each CS or IS beam is written as a separate file, and the data for a single beam may also be split into multiple 'parts', each containing a subset of the total number of subbands used. The filenames have the format: **LNNNNNN_SAPNNN_BNNN_SN_PNNN_bf.raw** (and the same with the extension **.h5** for the associated metadata), which contain the 6-digit generic LOFAR observation ID, the 3-digit sub-array pointing number (this is restricted to a maximum value of 488, and more typically only one or a few SAPs are used), the 3-digit beam number (this includes both IS and CS beams; these are ordered such that, for each SAP, first the manually-specified CS beams are listed, then the single IS beam, and then the CS beams specified via rings), the 1-digit Stokes parameter (0,1,2,3 = I,Q,U,V or XX,XY,YX,YY depending on whether raw voltages are being used or not) and a 3-digit part number (the subset of subbands that are included; the maximal split is one file per subband, but often all subbands are recorded to a single file in order to facilitate further offline analysis). The raw data itself is written as a binary file with the extension .raw. All of the associated header information is contained in a much smaller file of the same base name but extension .h5. Note the the raw data samples in the .raw file are 32-bit floats. There is a facility to convert these to 8-bit samples offline.

PSRFITS Format In order to facilitate data analysis with third party, pulsar-specific tools, the LOFAR HDF5 data is sometimes converted to the PSRFITS format, and may be archived as such (using the same base file name as the original .raw file, but with the new extension .fits). In this conversion, the data samples are repacked from 32-bit floats to 8-bit values. PSRFITS is the most common data standard in the pulsar astronomy community, and is described in Hotan et al. (2004) and van Straten et al. (2010). The primary 3rd party pulsar packages that are used for data analysis are `PRESTO` (http://www.cv.nrao.edu/~sransom/presto/; Ransom 2001), `dspsr` (http://dspsr.sourceforge.net/manuals/dspsr/; van Straten and Bailes 2011), and `PSRCHIVE` (http://psrchive.sourceforge.net/; Hotan et al. 2004; van Straten et al. 2010, 2012). Some of the utilities from these packages are described in more detail below. A tutorial on how to use the `PRESTO` package is available here: http://www.cv.nrao.edu/~sransom/PRESTO_search_tutorial.pdf.

LuMP and Other Formats Some of the standalone single station backends use other data formats internally, such as LuMP.

RFI Excision LOFAR BF data is significantly affected by RFI, but the situation is manageable—see Offringa et al. (2013) for a general description of the LOFAR RFI environment and Stappers et al. (2011) for some more specific comments on RFI at high time resolution. Much of the RFI is concentrated in certain narrow frequency bands that are almost always corrupted. There are also short-timescale bursts of RFI that need to be excised. Several tools exist for RFI excision. For example, on raw PSRFITS data, one can use the PRESTO tool `rfifind`, which automatically creates a time-frequency RFI mask based on the maximum value, standard deviation, and mean of each chunk of the data. Similar RFI excision tools from PSRCHIVE can also be used (`psrzap` or `pazi`). Typically, 5–20% of the data is excised in 80-MHz HBA observations (from 110 to 190 MHz). In the HBA band, there is not obviously a strong difference between night-time and daytime observing. At the lowest frequencies (10–40 MHz), there is definitely an advantage to observing between 2 and 5 am local time.

Dedispersion This can be done with `dspsr` or the PRESTO tool `(mpi)prepsubband`. For coherent dedispersion (for a detailed description, see Lorimer and Kramer 2004), `dspsr` can be used directly on the raw HDF5 data. Care should be taken at very low observing frequencies and relatively high dispersion measures because the dispersive delay can become quite extreme and may demand a large amount of compute memory. For example, for a dispersion measure of 50 pc cm^{-3} (still relatively low), the dispersive delay from 90 to 10 MHz is 2049 s (strongly dominated by the lowest observed frequency)! For comparison, the same signal would only be delayed by 11 s between 190 and 110 MHz. In any case, note that you can end up with problems if you try to dedisperse a data set that is not significantly longer than the dispersive delay across the observed bandwidth. For incoherent dedispersion, the PRESTO tool `(mpi)prepsubband` is often used. It is possible to generate multiple trial DMs simultaneously, which is needed for blind searches.

Timing For the purposes of pulsar timing, all LOFAR BF observations performed through the central system (i.e. not in standalone mode at an international station), *and regardless of which stations are actually used*, are phase-referenced to the center of CS002's LBA field, taken to be at (x, y, z) coordinates of (3826577.462 m, 461022.624 m, 5064892.526 m) in the ETRS89 system (Stappers et al. 2011). This consistent reference position was chosen in order to facilitate barycentering of the data using the standard 3rd party packages TEMPO (http://tempo.sourceforge.net/) and TEMPO2 (http://www.atnf.csiro.au/research/pulsar/tempo2/; Hobbs et al. 2006). LOFAR's position is included in these packages and the standard cite codes are "t" and "LOFAR" (or "t", "LF", "lofar" as aliases) for TEMPO and TEMPO2, respectively. If using data from a LOFAR international station, acquired in standalone mode, then take care that the station's position is recognized.

Single-Pulse Searches A simple search for dispersed single-pulses can be done using PRESTO's `single_pulse_search.py` tool. The input for this program are dedispersed timeseries in the PRESTO .dat format (which is actually just a flat binary file with 32-bit floats for the samples) and the accompanying .inf file, which contains the header information. Both of these files can be generated using PRESTO's `(mpi)prepsubband` tool. `single_pulse_search.py` uses a box-car match-filter to search the timeseries on a range of timescales (typically from the intrinsic sampling rate up to $\sim$100$\times$ this). Candidate single-pulse events are written to a .singlepulse file (in ASCII format). Multiple .singlepulse files from multiple DMs can be combined into a single diagnostic plot, also using `single_pulse_search.py`. For blind searches, there can be many thousands of DM trials, as well as multiple beams. As such, the number of detected events above the threshold (typically 5 sigma, but this can be set higher) is very large and clever sifting code is needed to identify the most promising astrophysical pulses (Coenen 2013).

Periodicity Searches Full-blown periodicity searches over the full astrophysically interesting DM space ($\mathrm{DM} = 0 - 1000\,\mathrm{pc\,cm^{-3}}$) require much larger amounts of computing power than is available on CEP4. To maintain high effective time resolution, the DM space must be searched in very small trial steps—typically in steps of $0.01\,\mathrm{pc\,cm^{-3}}$ in the HBA band, and an even finer grid in the LBA band. Each of these thousands of trials needs to be searched for a periodic signal using an FFT-based tool like `PRESTO`'s `accelsearch`. During this search, it is also possible to search over a possible drift of the signal in the Fourier power spectrum, which can be imparted by a Doppler shift of the signal in a binary orbit. This extra search dimension increases the computational cost even further.

Pulsar Folding For known sources, pulsar folding and dedispersion (optionally using an ephemeris) can be done with `dspsr` or `PRESTO`'s `prepfold` tool. Both of these methods can be used for pulsar timing though, when using `prepfold`, take care to use the *-timing* option (times-of-arrival can then later be generated using `get_TOAs.py`). `prepfold` is more routinely used for folding periodicity candidates in a search. `dspsr` is more commonly used for precision timing (because it offers coherent dedispersion). Both codes produce output data cubes (pulsar brightness as a function of time, frequency, and rotational phase) that can be further manipulated, refolded, and excised of RFI using a variety of third party tools.

Cyclic Spectroscopy More advanced signal processing techniques like cyclic spectroscopy (Demorest 2011; Archibald et al. 2014) often require access to the CV data. If such data is to be archived in the LTA, then it is best to first convert it to 8-bit samples. Tools to perform cyclic spectroscopy, which can provide simultaneous high frequency resolution and pulse-phase resolution, are available in `dspsr`, but one is cautioned that properly excising RFI is an important step in the process.

Deriving Rotation Measures RM-synthesis can be used to measure Faraday rotation. This method is somewhat implemented in `rmfit` in the `PSRCHIVE` suite—see Fig. 14.6 for an example of the output. However, more recently, an RM-synthesis routine has been written in Python, using the `PSRCHIVE` interface[6] to read, time/frequency average, and dedisperse the data. RM-synthesis is then performed using the Stokes Q and U parameters for pulse phase bins with total linear polarization greater than the specified S/N threshold (usually S/N > 7).

---

[6]See http://psrchive.sourceforge.net/manuals/python/.

Polarimetric and Flux Calibration LOFAR BF polarimetric data have been successfully calibrated both for instrumental effects (Noutsos et al. 2015) and the ionosphere (Sotomayor-Beltran et al. 2013). Likewise, a scheme exists to flux calibrate LOFAR pulsar observations (e.g. Kondratiev et al. 2015, in prep.). Manuals are in development to describe these procedures and codes. Interested users should contact Science Support for more information.

14.5 The LOFAR Standard Pulsar Pipeline (PulP)

The preceding section outlined many of the most commonly used tools for LOFAR pulsar data analysis. These tools are the building blocks of the LOFAR Standard Pulsar Pipeline (PulP). Recently, PulP has been incorporated into the MoM/Scheduler framework such that pipeline parameters can be specified beforehand and then scheduled for execution after the observation is complete. Depending on how it is configured, PulP can optionally:

- Fold/dedisperse the beams for a particular specified set of pulsars. It is also possible to let the pipeline search the catalog for known pulsars near the beam centers and to process these accordingly. There is a default repository of rotational ephemerides (i.e. 'parfiles'), but the pipeline can also be directed to specific parfiles. One can also run a limited search over the folding parameters to try and optimize the S/N by tweaking the folding period and DM. Extra options to `dspsr` and `prepfold` can also be specified.
- Create an `rfifind` mask for each of the beams.
- Convert the HDF5 .raw data to PSRFITS format.
- Convert the HDF5 .raw data from 32-bit samples to 8-bit samples.
- Create diagnostic summary plots of the reduced data (e.g., Fig. 14.3).
- Create a simple dynamic spectrum for each of the beams.
- Run a limited single-pulse search, which is useful in the case of observing intermittent pulsars or RRATs.
- Tailor whether processing is done on all of the available data products for a particular obsID, or whether processing is only performed on IS, CS, CS, or FE data products.

14.6 The Dynamic Spectrum Toolkit

For other science cases, analysis of the dynamic spectra is often desired. The Dynamic spectrum Toolkit Container (DTC) is a complete package of tools to reduce, process and visualise dynamic spectra from LOFAR beam-formed data:

- 'Quicklook' tool: This tool generates quick look dynamic spectra of the data to visualise them and determine which part(s) are useful for further processing.

Plots generated via this tool have a fixed time resolution of one second and frequency resolution of one sub-band.

- Rebinning tool: Extract particular time and frequency ranges of interest and rebin these to average in time and/or frequency.
- Subtraction tool: Subtract one beam from another (for example to subtract an off-source beam from an on-source) using Z=X−(k*Y) where k is a free scaling factor.
- Linear Polarisation tool: Convert I, Q, U, V dynamic spectra to I, linear, PA, Total dynamic spectra.
- Visualisation tool to explore the data.

Whilst DTC was primarily designed for use with LOFAR computing resources, both centrally and for international LOFAR stations being run in Local mode, it is also possible to install and use it on your own facilities. Full installation details are available from the LOFAR wiki page:

http://www.lofar.org/operations/doku.php?id=dysnpec_install.

All of the primary functions of the DTC package convert beam-formed data to, or use beam-formed data already stored, in a 'dynamic spectrum' HDF5 data format. For further details and tutorials, the reader is referred to the LOFAR wiki page:

http://www.lofar.org/operations/doku.php?id=dysnpec_tools

14.7 Future Prospects

Efforts are still underway to continue expanding the suite of BF observational functionality. These can both improve the quality of data delivered by existing modes, and expand the flexibility of their settings. It is useful to explicitly state some of the limitations in the current system and how these can be improved: (i) Every BF data beam must be synthesized by the same ensemble of stations. This is also the case for simultaneous BF+IM observations. Having separate station lists per beam/data type would enable a larger range of beam shapes/sizes to be used in a single observation. Having separate station lists for BF/IM data would allow better imaging quality by using the full Dutch array, while only using the Core for CS beam formation. Similarly, BF/IM data are required to use the same subband list, but tuning these separately would be preferable (e.g. IM data often skips subband numbers, while BF data almost always uses contiguous subbands). (ii) While BF (or IM) observations are running on a particular subset of stations (usually the Core stations) then the rest of the stations (usually Remote stations) remain idle. The possible exception is international stations running independently in standalone mode. Parallel observations would make much more efficient use of the LOFAR collecting area, only a fraction of which is used in typical observations. Once the current bottleneck of data processing on CEP4 (and beyond) is overcome, then this will become a very attractive feature. (iii) RFI excision can only be done on the output data stream, not on individual station streams, before online combination. Each station has its own RFI environment, and as such, removing RFI on a per-station basis can greatly improve the quality of the combined data sets. Also, BF observations are sensitive to malfunctioning stations (which cannot simply be flagged-out *post-facto*, as is done for IM data). (iv.) Currently the COBALT beam-former and correlator can only output samples as 32-bit floats. Re-bitting the streaming COBALT output to 8, 4 or 2-bit samples would enable larger number of beams within the same data rate.

Acknowledgements Many thanks to Vlad Kondratiev and Anastasia Alexov for being key contributors in the development and testing of the LOFAR Standard Pulsar Pipeine (PulP; `pulp.py`). Along with the author (JH) of this chapter, Ben Stappers is co-lead of the LOFAR Pulsar Working Group, and has played a large role in the development of LOFAR's BF modes. Anya Bilous provided the lovely plots for Fig. 14.5. Many thanks also to Charlotte Sobey and Aris Noutsos for providing descriptions of LOFAR RM synthesis and polarimetric calibration. Also many thanks to the 20-odd members of the LOFAR Pulsar Working Group for the great contributions that they have made to LOFAR BF commissioning and science exploitation. Many thanks to Nicolas Vilchez for his single-handed development of the Dynamic Spectrum Toolkit. Jason Hessels gratefully acknowledges funding from an NWO Vidi fellowship and ERC Starting Grant DRAGNET (337062).

References

Alexov, A., Hessels, J., Mol, J.D., Stappers, B., van Leeuwen, J.: In: Mizumoto, Y., Morita, K.-I., Ohishi, M. (eds.) Astronomical Data Analysis Software and Systems XIX. Astronomical Society of the Pacific Conference Series, vol. 434, p. 193 (2010)
Alexov, A., Schellart, P., ter Veen, S., et al.: In: Ballester, P., Egret, D., Lorente, N.P.F. (eds.) Astronomical Data Analysis Software and Systems XXI. Astronomical Society of the Pacific Conference Series, vol. 461, p. 283 (2012)
Archibald, A.M., Kondratiev, V.I., Hessels, J.W.T., Stinebring, D.R.: Astrophys. J. **790**, L22 (2014)
Backer, D.C., Kulkarni, S.R., Heiles, C., Davis, M.M., Goss, W.M.: Nature **300**, 615 (1982)
Bassa, C.G., Patruno, A., Hessels, J.W.T., et al.: Mon. Not. R. Astron. Soc. **441**, 1825 (2014)
Bilous, A., Hessels, J., Kondratiev, V., et al. (2014). arXiv e-prints
Coenen, T.: PhD thesis, Anton Pannekoek Instituut, Universiteit van Amsterdam, Postbus 94249, Amsterdam, The Netherlands (2013)
Coenen, T., van Leeuwen, J., Hessels, J.W.T., et al.: Astron. Astrophys. **570**, A60 (2014)
Demorest, P.B.: Mon. Not. R. Astron. Soc. **416**, 2821 (2011)
Dolch, T., Lam, M.T., Cordes, J., et al.: Astrophys. J. **794**, 21 (2014)
Fallows, R.A., Asgekar, A., Bisi, M.M., Breen, A.R., ter Veen, S.: Sol. Phys. **285**, 127 (2013)
Fallows, R., Coles, W., McKay-Bukowski, D., et al.: J. Geophys. Res. Space Phys. **119**, 10544–10560 (2014)
Hassall, T.E., Stappers, B.W., Hessels, J.W.T., et al.: Astron. Astrophys. **543**, A66 (2012)
Hassall, T.E., Stappers, B.W., Weltevrede, P., et al.: Astron. Astrophys. **552**, A61 (2013)
Hermsen, W., Hessels, J.W.T., Kuiper, L., et al.: Science **339**, 436 (2013)
Hewish, A., Bell, S.J., Pilkington, J.D.H., Scott, P.F., Collins, R.A.: Nature **217**, 709 (1968)
Hobbs, G.B., Edwards, R.T., Manchester, R.N.: Mon. Not. R. Astron. Soc. **369**, 655 (2006)
Hotan, A.W., van Straten, W., Manchester, R.N.: Publ. Astron. Soc. Aust. **21**, 302 (2004)
Karako-Argaman, C., Kaspi, V.M., Lynch, R.S., et al.: (2015). arXiv e-prints
Kondratiev, V.I., Verbiest, J.P.W., Hessels, J.W.T., et al.: Astron. Astrophys. **585**, A128 (2016)
Kramer, M., Lyne, A.G., O'Brien, J.T., Jordan, C.A., Lorimer, D.R.: Science **312**, 549 (2006)

Lorimer, D.R., Kramer, M.: Handbook of Pulsar Astronomy. Cambridge Observing Handbooks for Research Astronomers, vol. 4. Cambridge University Press, Cambridge (2004)
Lorimer, D.R., Bailes, M., McLaughlin, M.A., Narkevic, D.J., Crawford, F.: Science **318**, 777 (2007)
Lyne, A., Hobbs, G., Kramer, M., Stairs, I., Stappers, B.: Science **329**, 408 (2010)
McKay-Bukowski, D., Vierinen, J.-P., Virtanen, I., et al.: IEEE Trans. Geosci. Remote Sens. **53**, 1440 (2014)
McLaughlin, M.A., Lyne, A.G., Lorimer, D.R., et al.: Nature **439**, 817 (2006)
Mol, J.D., Romein, J.W.: (2011). arXiv e-prints
Morosan, D.E., Gallagher, P.T., Zucca, P., et al.: Astron. Astrophys. **568**, A67 (2014)
Noutsos, A., Sobey, C., Kondratiev, V.I., et al.: (2015). arXiv e-prints
Offringa, A.R., de Bruyn, A.G., Zaroubi, S., et al.: Astron. Astrophys. **549**, A11 (2013)
Papitto, A., Ferrigno, C., Bozzo, E., et al.: Nature **501**, 517 (2013)
Pilia, M., Hessels, J.W.T., Stappers, B.W., et al.: Astron. Astrophys. **586**, A92 (2016)
Ransom, S.M.: PhD thesis, Harvard University (2001)
Rickett, B.J.: Annu. Rev. Astron. Astrophys. **28**, 561 (1990)
Serylak, M., Karastergiou, A., Williams, C., et al.: In: van Leeuwen, J. (ed.) IAU Symposium, vol. 291, pp. 492–494 (2013)
Sobey, C., Young, N.J., Hessels, J.W.T., et al.: (2015). arXiv e-prints
Sotomayor-Beltran, C., Sobey, C., Hessels, J.W.T., et al.: Astron. Astrophys. **552**, A58 (2013)
Spitler, L.G., Cordes, J.M., Hessels, J.W.T., et al.: Astrophys. J. **790**, 101 (2014)
Stappers, B.W., Hessels, J.W.T., Alexov, A., et al.: Astron. Astrophys. **530**, A80 (2011)
Stappers, B.W., Archibald, A.M., Hessels, J.W.T., et al.: Astrophys. J. **790**, 39 (2014)
Stovall, K., Lynch, R.S., Ransom, S.M., et al.: Astrophys. J. **791**, 67 (2014)
Thornton, D., Stappers, B., Bailes, M., et al.: Science **341**, 53 (2013)
van Haarlem, M.P., Wise, M.W., Gunst, A.W., et al.: Astron. Astrophys. **556**, A2 (2013)
van Straten, W., Bailes, M.: Publ. Astron. Soc. Aust. **28**, 1 (2011)
van Straten, W., Manchester, R.N., Johnston, S., Reynolds, J.E.: Publ. Astron. Soc. Aust. **27**, 104 (2010)
van Straten, W., Demorest, P., Oslowski, S.: Astron. Res. Technol. **9**, 237 (2012)

Index

A

Active galactic nucleus (AGN), 4, 14, 179
Air showers, 20, 212, 213, 217, 219–222
Aliasing, 21, 45, 125, 127, 129, 130
Alternating direction implicit (ADI), 153
Amplifier, 20, 37–39, 45, 46, 66, 67, 166, 215
Amsterdam-ASTRON Radio Transients Facility and Analysis Center (AARTFAAC), 41, 46
Analog to Digital Converter (ADC), 45
Analogue beam, 33, 78, 220, 228
AOflagger, 60, 62, 143, 204
Aperture array, 70, 118, 141, 174, 177
Aperture synthesis interferometry, 9
Aperture tile in focus (APERTIF), 14
A-projection, 96–97, 110, 111, 118, 130, 132–133, 147
Arecibo, 8
Array beam, 24–26, 28–35. 78, 177, 227–229
Astrometry, 75
ASTRON, 9, 26
Astronomical image processing system (AIPS), 172, 187, 189, 196, 198, 199, 206–208
A-team, 60–62, 141, 143, 144, 151, 183
A-term, 111, 132, 133, 147
Australia Telescope Compact Array (ATCA), 78

B

Balun, 20
Bandpass, 27, 28, 45, 67, 79–80, 97, 145, 184, 187, 203–209
Bandwidth smearing, 97, 151
Baseline, 10, 22, 24, 28, 32, 41, 59–62, 65, 68, 69, 72, 73, 75, 76, 86, 99, 105, 106, 110, 112, 113, 118–120, 124, 131, 133, 143, 144, 149, 160, 166, 172, 174, 179–199, 216, 228, 230, 231, 235
Bayesian inference, 84
Beamformer (BF), 22, 24, 26, 28–30, 33, 44, 47, 48, 78, 220, 227–230, 232, 238, 239, 241–243
Beamlet processor (BLP), 47–48
Beamlets, 47–49, 228, 229
Beamlet statistics (BST), 48–49
Black Board Self-Calibration (BBS), 81–83, 87, 106, 113, 114, 135, 143, 145–148, 150, 152, 153, 177, 198, 199, 205, 207
Bootstrapping, 79, 80

C

Celestial equator, 141, 148
Central processing (CEP), 25, 26, 42, 46, 49, 212, 213
CEP3, 25
CEP4, 25, 27, 35, 232, 233, 240, 242
Circular polarization, 161, 162
CLEAN, 11, 83, 94, 95, 113, 121–123, 134, 153, 166, 170, 208, 216
Clock, 21, 22, 24, 26, 30, 34, 42–45, 49–51, 88, 99, 107–110, 160, 184, 186, 188, 198, 203, 205, 226, 228, 235, 237
Closure, 68
Coherence, 81, 141
Coherency matrix, 165, 174
Coherent dedispersion, 233, 234, 239, 240

G. Heald et al. (eds.), *Low Frequency Radio Astronomy and the LOFAR Observatory*, Astrophysics and Space Science Library 426,
https://doi.org/10.1007/978-3-319-23434-2

Common Astronomy Software Applications (CASA), 56, 58–60, 118, 128, 147, 153, 189, 199, 206–208
Complex aperture, 34
Complex voltage (CV) mode, 233, 237
Compressed sensing, 83–84
Concentrator node, 51
Continuum subtraction, 207–208
Convolution theorem, 123, 124
Core, 19, 22–24, 34, 41, 42, 51, 59, 71–73, 75, 76, 78, 85, 86, 99, 107, 109, 110, 123, 144, 177, 180–182, 184, 186, 188, 191, 192, 195, 198, 220, 221, 228–232, 235–237, 242
Correlator, 10, 20, 26–28, 41, 42, 56, 66, 143, 144, 163–166, 173, 184, 186, 188, 192, 203, 206, 212, 228, 229, 232, 243
Correlator and Beam former Application for the LOFAR Telescope (COBALT), 26–29, 192, 228, 232, 233, 235, 236, 243
Cosmic ray (CR), 11, 13, 20, 41, 212, 213, 216–220, 222, 226
Cost function, 121, 122, 134
Crosslet Statistics (XST), 49
Cyclic spectroscopy, 234, 235, 240

D

Data Access Library (DAL), 238
Deconvolution, 74, 83, 94, 96, 111, 118, 133, 134, 153
Dedispersion, 233, 234, 239, 240
Default Pre-Processing Pipeline (DPPP), 55–57, 59, 61, 62, 134, 136
Degridding, 118, 122–130, 134
Delay, 10, 24, 26–28, 30, 39, 40, 44, 46–48, 51, 60, 65–69, 77, 81, 99, 103, 105–108, 166, 167, 172, 177, 180, 186–192, 194–196, 198, 205, 229, 230, 234, 235, 239
Demixing, 55–62, 141, 143, 144, 151, 152, 204
Differential Faraday rotation, 111–113, 172–174, 189, 198
Differential refraction, 76, 77, 105, 111
Digital beam, 15, 30, 33, 34, 174, 175, 177, 228
Digital beam forming, 15
Dipole, 7, 8, 15, 21, 24, 33, 34, 38, 39, 41, 44, 46, 78, 174–177, 213–215, 247
Dipole beam, 33
Dirac delta function, 34, 123, 126, 130
Direct Fourier Transform (DFT), 83, 122, 123
Directional calibration, 60, 61, 143, 151, 184
Direction cosines, 119
Direction-dependent, 86, 87, 99, 101, 147, 152–154
Direction-dependent calibration, 87, 99, 101, 152–154
Dirty beam, 10, 11, 94
Dispersion, 169, 206, 227, 233–235, 237, 239, 240
Dispersive delay, 105–111, 188, 189, 234, 239
Doppler correction, 208
dspsr, 238–241
Dwingeloo, 8
Dynamic range, 70, 71, 76, 81, 86–89, 97, 140, 141, 144, 184
Dynamic spectrum Toolkit Container (DTC), 241, 242

E

Effelsberg, 8
Element beam, 30, 33, 34, 78, 141, 166, 175–177
Elliptical polarization, 160, 161
Epoch of Reionization (EoR), 11–13, 171
Evolutionary Map of the Universe (EMU), 14
Exloo, 19, 41

F

Facet calibration, 87–89, 152, 154
Facets, 88, 131, 154
Faraday depth, 168, 169, 171
Faraday dispersion function, 169
Faraday rotation, 12, 41, 105, 111–115, 159, 166–174, 177, 189, 198, 227, 233, 235
Faraday spectrum, 169
Faraday thickness, 168
Fast Fourier Transform (FFT), 47, 123, 129–131, 216, 240
Fast Radio Burst (FRB), 14, 233
F-, D-and E-layers of the ionosphere, 103
Field based calibration, 87
Field Programmable Gate Array (FPGA), 46
Fine delay compensation, 27, 28
First In First Out (FIFO) buffer, 46
Flagging, 25, 28, 55–62, 82, 85–86, 141–144, 199, 204, 208–209
Flux scale, 79, 80, 94, 99, 145, 193
Fly's-Eye (FE) beam, 229, 231
Fourier transform, 10, 30, 34, 71, 82–84, 93, 94, 96, 97, 120, 123–125, 127, 129–132, 163, 164, 170, 171, 216, 217

Fresnel scale, 75, 105
FRING, 187, 189, 190, 198
Fringe-fitting, 186, 187, 189, 191, 194, 196, 198

G
Gain, 15, 22, 30–34, 39, 45, 56, 66, 67, 72, 73, 78–80, 84, 105, 106, 110, 128, 133, 145–148, 150, 154, 166, 174, 177, 186, 190, 192, 213–215
Galactic and Extragalactic All-sky Murchison Widefield Array (GLEAM), 75, 83FI
Galaxy cluster, 4, 13
Geometric delay, 30, 65, 66, 81
Giant Metrewave Radio Telescope (GMRT), 9, 12, 15, 75, 87, 99, 147
Global calibration, 87
Global Positioning System (GPS), 50, 51, 104, 105, 109, 111, 113, 114, 171, 172, 188
Global Sky Model (GSM), 147, 150
Graphical Processing Unit (GPU), 26, 28, 29
Grating response, 31–33
Green Bank Telescope (GBT), 8
Gridding, 118, 122–128, 130, 132–136, 153
Ground plane, 33, 38, 39, 175, 176

H
Hermitian, 68, 121
Hierarchical Data Format 5 (HDF5), 29, 238, 239, 241, 242
High Band Antenna (HBA), 12, 15, 19–22, 24, 28, 30–33, 35, 38–45, 59, 60, 71, 75, 78, 86, 99, 101, 106, 108–112, 117, 121, 140, 142–147, 151, 168, 172, 174, 177–179, 183, 186, 191, 202–206, 208, 209, 212, 220, 221, 228–230, 234, 237, 239, 240
High pass filter, 22, 45
Hilbert envelope, 217, 218, 221
Hilbert transform, 217–219

I
Image Domain Gridding (IDG), 153
Incoherent Stokes Beam (IS), 229, 231
Instrumental polarization, 172, 173
Intergalactic medium (IGM), 12, 233
Interleaved observation, 84
International LOFAR Telescope (ILT), 22, 23, 41, 233
International stations, 22–24, 41, 42, 49, 76, 179–183, 185–193, 198, 233, 242
Interstellar medium (ISM), 3, 4, 168, 171, 202, 209, 227, 233
Intracluster medium (ICM), 4
Ionosonde, 104
Ionosphere, 15, 34, 74–77, 82, 88, 99, 103–107, 110–115, 160, 171–174, 186–188, 190, 227, 229, 235, 241
Ionospheric phasescreen, 77, 110
Ionospheric refraction, 76, 77
Isoplanatic patch, 184, 205

J
Jodrell Bank, 8
Jones matrix, 67, 165, 171, 173, 177, 215

K
Kolmogorov turbulence, 105

L
Leakage, 67, 112, 166, 174–177
Levenberg-Marquardt (LM) least-squares solver, 152
Linear polarization, 105, 111, 160, 162, 240
LOFAR Radboud air shower Array (LORA), 41, 213, 220, 221
Long-Baseline Calibrator Survey (LBCS), 197
Long term archive (LTA), 20, 25, 240
Long Wavelength Array (LWA), 8, 15, 201
LoSoTo, 108, 153
Low band antenna (LBA), 20, 38, 179, 202, 204, 213, 228
Low Frequency Array (LOFAR), 5, 8, 9, 11–16, 19–51, 55–62, 65–90, 93–101, 103–108, 110, 111, 117, 118, 121, 131, 139–154, 159–199, 201–209, 211–222, 225–243
Low Frequency Array Cassiopeia A Spectral Survey (LCASS), 204
Low Frequency Array imaging Cookbook, 57, 81, 84, 118, 185, 195
Low Frequency Array Two-metre Sky Survey (LoTSS), 13, 140, 149
Low Noise Amplifier (LNA), 37–39
Low pass filter, 22
LSMTool, 153

M
Magnetic field, 4, 11, 12, 38, 70, 105, 111, 113–115, 168, 171, 220, 227
Magnetism, 5, 11, 12
Major cycle, 151, 153

Management of Measurements (MOM), 232, 241
Manchester encoding, 44
Measurement equation, 67, 68, 72, 86, 87, 112, 118, 120, 122, 130, 132, 134, 145, 166, 174
Measurement set (MS), 28, 78, 82, 135, 198, 199, 203, 205
Mixing matrix, 61
Monitoring and Control (MAC), 43, 50
Mosaic, 88, 145, 149, 150, 154, 227
Mueller matrix, 112
Multifrequency Snapshot Sky Survey (MSSS), 76, 77, 83, 144, 147, 149, 150, 192, 197
Multifrequency synthesis (MFS), 95, 181
Multi-scale clean, 94, 95
Murchison Radio-astronomy Observatory (MRO), 16
Murchison Widefield Array (MWA), 8, 13, 15, 75, 201

N
NE2001, 197
Non-dispersive delay, 188, 189, 194
NRAO VLA Sky Survey (NVSS), 147
NuMoon, 213
Nyquist Theorem, 21
Nyquist zone, 21, 22, 30

O
Online Application Processing (OLAP), 82
Optical depth, 207, 208
Oscillating tiles, 232, 237

P
Parallactic angle, 166
Parkes, 8
Parmdb, 82
Parset, 56, 57, 62, 82, 136
Peeling, 87, 97, 99, 100, 111, 154
Phased array, 5, 8, 15, 19, 29–35, 78, 159
Phase screen, 77, 110, 111, 154
Pierce-points, 110
Pierre Auger observatory, 222
Piggy-back mode, 220
Planetary emission, 226
Plasma frequency, 106
Point spread function (PSF), 94, 122, 169, 185, 230
Polarimetry, 159, 160, 168, 177, 178, 227
Poly phase filter (PPF), 27, 29, 206
P/O ratio, 152
Precision Array for Probing the Epoch of Reionization (PAPER), 13, 15
PRESTO, 238–240
Primary beam, 73, 78, 96, 99, 117, 119, 122, 132, 148–150
Primary calibration, 141, 145, 147
Prolate Spheroidal Wave Function (PSWF), 128, 129, 132
PSRCHIVE, 238–240
PSRFITS, 238, 239, 241
Pulsar, 24, 34, 36, 171, 172, 209, 225, 227, 230, 231, 233–241
Pulsar Pipeline (PulP), 236, 241
Pulse Width Modulator (PWM), 44
PyBDSF, 150, 153
Python Blob Detector and Source Finder (PyBDSF), 150, 153

R
Radio frequency interference (RFI), 4, 6, 45, 55–62, 85, 94, 142, 143, 208, 214, 216–218, 221, 229, 231, 237–240, 242, 243
Radio frequency (RF) transformer, 21, 38, 39
Radio Interferometric Measurement Equation (RIME), 67, 145
Radio recombination line (RRL), 4, 202–204, 208, 209
Receiver unit (RCU), 21, 38, 43–48, 211
Remote Station Processor (RSP), 43–46, 49
Ring splitter, 42
rl phase difference, 161, 167
RMextract, 114
Rotation measure (RM), 111, 113–115, 167–173, 236, 240
Rotation measure (RM) clean, 170
Rotation measure (RM) cube, 171
Rotation measure (RM)-selfcal, 172, 173
Rotation measure (RM) Spectrum, 169
Rotation measure Spread Function (RMSF), 169, 170
Rotation measure (RM)-synthesis, 240

S
Scattering, 197, 233–235
Scheduler, 232
Scintillation, 105, 106, 227, 232–234
Secondary calibrator, 191, 194, 195
Self-calibration, 33, 70–79, 81, 82, 84, 85, 88, 101, 107, 150–154, 172, 177, 186, 209

Shift+averaging procedure, 196–197
Sky noise, 37–39
Slant TEC, 104
Smart demix, 61–62, 141, 144, 151
Smart demixing, 61, 141, 144, 151
Smearing, 97, 144, 151, 183–185, 192, 234
Snapshots, 75, 76, 131, 144, 147, 195, 197
Solar bursts, 226
Solution interval, 72–74, 85, 86, 189
Source Peeling and Atmospheric Modeling (SPAM), 154
Space Alternating Generalised Expectation Maximisation calibration (SAGECal), 89, 143
Space Weather, 11, 13
Spectral energy distribution (SED), 79, 80, 204
Spectral index, 70, 83, 95, 197, 207
Square Kilometre Array (SKA), 11, 14, 16, 87
Standard Imaging Pipeline (SIP), 139–141, 144, 145, 147, 150–153
Station-bandpass correction, 27, 28
Station calibration tables, 46
Station central oscillator (SCO), 43, 49–51
Stokes parameters, 28, 161–165, 175–177, 218, 219, 232, 238
Sub-array pointing (SAP), 228, 229, 231, 237, 238
Subband, 22, 25–30, 46–49, 82, 106, 142–145, 147, 148, 151, 168, 169, 183–185, 196, 199, 203–209, 211, 212, 226, 228, 229, 232–234, 237–239, 242
Subband statistics (SST), 48, 49
Subrack, 43
Summator, 39, 40, 44
Sun, 5, 11, 13, 60, 170, 188, 208, 227
Supernova remnants, 179, 203, 219
Superterp, 13, 22, 34, 41, 86, 145–148, 213, 214, 221, 229, 230, 235
Support, 20, 22, 26, 83, 127, 128, 131, 208, 209, 213, 237
Surveys, 6, 7, 11, 13–15, 22, 34, 72, 75, 76, 79, 80, 83, 99, 140, 144, 145, 147, 149, 150, 197, 204, 227, 231, 236
SWARP, 149
Synchrotron radiation, 3, 12, 172
Syncoptics, 50, 51
Synthesized beam, 56, 180, 184
System equivalent flux density (SEFD), 189, 191

T

Table Query Language (TAQL), 198
Thermal noise, 84–86, 88, 121, 134, 140, 151, 183
Thermal radiation, 3, 13
Thin screen, 104
Tied array beam (TAB), 24–26, 28–29, 34, 35, 177, 227–229, 236
Tied array beam forming, 24, 25, 28, 177
Tied station, 78, 195, 198
TIFR GMRT Sky Survey Alternative Data Release (TGSS-ADR1), 75, 76, 83, 99, 147, 197
Tile, 15, 20, 21, 24, 30–35, 39, 41, 42, 44, 78, 159, 191, 220, 228–230, 232, 237
Tile beam, 33, 34, 39, 78, 159, 228–230
Time-average smearing, 97
Time Distribution System (TDS), 43, 50, 51
Total electron content (TEC), 76, 77, 88, 104–106, 108–111, 113–115, 154, 171–173, 188
Transient Buffer Board (TBB), 41, 43, 46, 212–214, 226
Transients, 11, 14, 41, 70, 212, 213, 226, 227
Traveling ionospheric disturbance (TID), 105
Troposphere, 3, 66
Two element interferometer, 65, 66

U

Ultrahigh energy cosmic ray (UHECR), 13
UV coverage, 24, 70, 75, 181, 183, 185
UVFITS, 196, 198
UV plane, 124, 126, 127, 149, 164, 207
UV tracks, 124, 131

V

Vertical TEC (VTEC), 76, 104
Very Large Array (VLA), 9, 15, 22, 78, 87, 101, 110, 147
Very Large Array Low-frequency Sky Survey (VLSSr), 75, 83, 99, 147, 197
Very Long Baseline Interferometry (VLBI), 174, 179, 184, 186, 188, 189, 191, 192, 194, 195, 198
Visibilities, 55–62, 67–69, 79, 80, 82–85, 87, 96, 97, 110, 118, 120–123, 131–134, 136, 139, 140, 143, 147, 152, 154, 163–165, 171, 175, 181, 183, 186, 192, 198, 226
Visibility, 10, 68, 74, 79, 84, 94, 121, 123, 124, 126, 127, 131–134, 139, 143, 150, 180, 183, 186, 196
VLA Sky Survey (VLASS), 14
Voronoi Tessellation, 154

W

WENSS, 75, 83, 147, 197
Westerbork Northern Sky Survey (WENSS), 75, 83, 147, 197
Westerbork Synthesis Radio Telescope (WSRT), 9, 74, 78, 87, 167, 172
Wirtinger derivatives, 87, 153
W-projection, 96–97, 117, 118, 122, 130–131, 147
Wsclean, 77, 96, 131
W-stacking, 131, 132

Astronomical Objects and Fields Index

Numbers & Symbols
3C84, 192, 194
3C196, 109, 192, 194
3C223, 36
3C244.1, 95, 97–10
3C295, 145

B
B0809+74, 36
B0823+26, 228
B0943+10, 227
B1245+67, 74
B1919+21, 230
B2021+51, 235, 236
B2111+46, 234

C
Cassiopeia A (CasA), 62, 78, 79, 184, 203, 204
Cygnus A (CygA), 9, 62, 72, 73, 78, 84, 86, 184, 203
Centaurus A (CenA), 9

E
ELAIS-N1, 114

H
HerA, 60
Hydra A (HydA), 60, 61

J
J0958+6533, 187, 190, 192, 193
J1810+1744, 234
J1819+3845, 71
Jupiter, 227

L
Lockman Hole, 97, 98, 100

M
M82, 179, 180, 183
Messier 51 (M51), 36
Milky Way, 6, 7, 12, 14

S
Sun, 5, 11, 13, 36, 188, 208

T
TauA, 60

V
Virgo A (VirA), 9, 36, 60

G. Heald et al. (eds.), *Low Frequency Radio Astronomy and the LOFAR Observatory*, Astrophysics and Space Science Library 426,
https://doi.org/10.1007/978-3-319-23434-2

The manufacturer's authorised representative in the EU is Springer Nature Customer Service Centre GmbH, Europaplatz 3, 69115 Heidelberg, Germany. If you have any concerns regarding our products, please contact ProductSafety@springernature.com

Printed and bound by CPI Group (UK) Ltd, Croydon, CR0 4YY

05/10/2026

02228138-0012